T0186297

International Journal of
Human–Computer Interaction

Vol. 13, No. 4   2001

Special Issue: Usability Evaluation
Guest Editor: James R. Lewis

Contents

INTERNATIONAL JOURNAL OF HUMAN–COMPUTER INTERACTION, *13*(4), 343–349
Copyright © 2001, Lawrence Erlbaum Associates, Inc.

Introduction:
Current Issues in Usability Evaluation

James R. Lewis
IBM Corporation

In this introduction to the special issue of the *International Journal of Human–Computer Interaction*, I discuss some current topics in usability evaluation and indicate how the contributions to the issue relate to these topics. The contributions cover a wide range of topics in usability evaluation, including a discussion of usability science, how to evaluate usability evaluation methods, the effect and control of certain biases in the selection of evaluative tasks, a lack of reliability in problem detection across evaluators, how to adjust estimates of problem-discovery rates computed from small samples, and the effects of perception of hedonic and ergonomic quality on user ratings of a product's appeal.

1. INTRODUCTION

1.1. Acknowledgements

I start by thanking Gavriel Salvendy for the opportunity to edit this special issue on usability. I have never done anything quite like this before, and it was both more difficult and more rewarding than I expected. The most rewarding aspect was working with the contributing authors and reviewers who volunteered their time and effort to the contents of this issue. The reviewers were Robert Mack (IBM T. J. Watson Research Center), Richard Cordes (IBM Human Factors Raleigh), and Gerard Hollemans (Philips Research Laboratories Eindhoven). Several of the contributing authors have indicated to me their appreciation for the quality of the reviews that they received. To this I add my sincere appreciation. Providing a comprehensive, critical review of an article is demanding work that advances the literature but provides almost no personal benefit to the reviewer (who typically remains anonymous). Doing this work is truly the mark of the dedicated, selfless professional.

1.2. State of the Art in Usability Evaluation

I joined IBM as a usability practitioner in 1981 after getting a master's degree in engineering psychology. At that time, the standard practice in our product develop-

Send requests for reprints to James R. Lewis, IBM Corporation, 8051 Congress Avenue, Suite 2227, Boca Raton, FL 33487. E-mail: jimlewis@us.ibm.com

ment laboratory was to conduct scenario-based usability studies for diagnosing and correcting usability problems. Part of the process of setting up such a study was for one or more of the usability professionals in the laboratory to examine the product (its hardware, software, and documentation) to identify potential problem areas and to develop scenarios for the usability study. Any obvious usability problems uncovered during this initial step went to Development for correction before running the usability study.

Alphonse Chapanis had consulted with IBM on the use of this rather straightforward approach to usability evaluation. Regarding an appropriate sample size, he recommended 6 participants because experience had shown that after watching about 6 people perform a set of tasks, it was relatively rare to observe the occurrence of additional important usability problems (Chapanis, personal communication, 1979).

A clear problem with this approach was that to run the usability study you needed a working version of the product. Since I first started as a usability practitioner, there have been two important lines of usability research that have affected the field. One is the development of additional empirical and rational usability evaluation methods (e.g., think aloud, heuristic evaluation, cognitive walkthrough, GOMS, QUIS, SUMI), many of which were motivated by the need to start initial usability evaluation earlier in the development cycle. The other is the comparative evaluation of these usability methods (their reliability, validity, and relative efficiency). The last 20 years have certainly seen the introduction of more usability evaluation tools for the practitioner's toolbox and some consensus (and still some debate) on the conditions under which to use the various tools. We have mathematical models for the estimation of appropriate sample sizes for problem-discovery usability evaluations (rather than a simple appeal to experience; Lewis, 1982, 1994; Nielsen & Landauer, 1993; Virzi, 1990, 1992). Within the last 12 years usability researchers have published a variety of usability questionnaires with documented psychometric properties (e.g., Chin, Diehl, & Norman, 1988; Kirakowski & Corbett, 1993; Lewis, 1995, 1999).

Yet many questions remain. For example

• Do we really understand the differences among the various usability evaluation methods in common use by practitioners? Do we have a good idea about how to compare usability evaluation methods?

• How do the tasks selected for a scenario-based usability evaluation affect the outcome of the evaluation? How do we decide what tasks to include in an evaluation and what tasks to exclude?

• How does the implicit assumption that we only ask participants to do tasks that are possible with a system affect their performance in a usability evaluation?

• Usability evaluation based on the observation of participants completing tasks with a system is the golden standard for usability evaluation, with other approaches generally considered discounted by one criterion or another. Usability practitioners assume that the same usability problems uncovered by one laboratory would, for the most part, be discovered if evaluated in another laboratory. To what extent do they know that this assumption is true? What are the implications for the state of the art if the assumption is not true?

- Sample size estimation for usability evaluations depends on having an estimate of the rate of problem discovery (p) across participants (or, in the case of heuristic evaluation, evaluators). It turns out, though, that estimates of this rate based on small-sample usability studies are necessarily inflated (Hertzum & Jacobsen, this issue). Is there any way to adjust for this bias so practitioners can use small-sample estimates of p to develop realistic estimates of the true problem-discovery rate and thereby estimate accurate sample size requirements for their studies?

- Is usability enough to assure the success of commercial products in a competitive marketplace? To what extent is "likeability" or "appealingness" affected by or independent of usability? How can we begin to measure this attribute?

- We feel like we know one when we see one, but what is the real definition of a *usability problem*? What is the appropriate level at which to record usability problems?

2. CONTRIBUTIONS TO THIS ISSUE

The contributions to this issue do not answer all of these questions, but they do address a substantial number of them.

2.1. "Usability Science. I: Foundations"

In this first article of the issue, Gillan and Bias describe the emerging discipline of usability science. The development and qualification of methods for usability design and evaluation require a scientific approach, yet they may well be distinct from other similar disciplines such as human factors engineering or human–computer interaction. In their article, Gillan and Bias reach across various modern disciplines and into the history of psychology to develop their arguments.

2.2. "Criteria for Evaluating Usability Evaluation Methods"

Hartson, Andre, and Williges (this issue) tackle the problem of how to compare usability evaluation methods. The presence of their article is additional testimony to the importance of the recent critique by Gray and Salzman (1998) on potentially misleading research ("damaged goods") in the current literature of studies that compare usability evaluation methods. Developing reliable and valid comparisons of usability evaluation methods is a far from trivial problem, and Hartson et al. lay out a number of fundamental issues on how to measure and compare the outputs of different types of usability evaluation methods.

2.3. "Task-Selection Bias: A Case for User-Defined Tasks"

Cordes (this issue) provides evidence that participants in laboratory-based usability evaluations assume that the tasks that evaluators ask them to perform must be possible and that manipulations to bring this assumption into doubt have dramatic effects on a study's quantitative usability measures (a strong bias of which many usability practitioners are very likely unaware). Cordes discusses how introducing user-defined tasks into an evaluation can help control for this bias and presents techniques for dealing with the methodological and practical consequences of including user-defined tasks in a usability evaluation.

2.4. "Evaluator Effect: A Chilling Fact About Usability Evaluation Methods"

This title of the article by Hertzum and Jacobsen (this issue) might seem a bit extreme, but the evidence they present is indeed chilling. Their research indicates that the most widely used usability evaluation methods suffer from a substantial evaluator effect—that the set of usability problems uncovered by one observer often bears little resemblance to the sets described by other observers evaluating the same interface. They discuss the conditions that affect the magnitude of the evaluator effect and provide recommendations for reducing it.

For me, this is the most disturbing article in the issue, in part because other investigators have recently reported similar findings (Molich et al., 1998). The possibility that usability practitioners might be engaging in self-deception regarding the reliability of their problem-discovery methods is reminiscent of clinical psychologists who apply untested evaluative techniques (such as projective tests), continuing to have faith in their methods despite experimental evidence to the contrary (Lilienfeld, Wood, & Garb, 2000). Although we might not like the results of Hertzum and Jacobsen (this issue), we need to understand them and their implications for how we do what we do as usability practitioners.

Often usability practitioners only have a single opportunity to evaluate an interface, so there is no way to determine if their usability interventions have really improved an interface. In my own experience though, when I have conducted a standard scenario-based, problem-discovery usability evaluation with one observer watching multiple participants complete tasks with an interface and have done so in an iterative fashion, the measurements across iterations consistently indicate a substantial and statistically reliable improvement in usability. This leads me to believe that, despite the potential existence of a substantial evaluator effect, the application of usability evaluation methods (at least, methods that involve the observation of participants performing tasks with a product under development) can result in improved usability (e.g., see Lewis, 1996). An important task for future research in the evaluator effect will be to reconcile this effect with the apparent reality of usability improvement achieved through iterative application of usability evaluation methods.

2.5. "Evaluation of Procedures for Adjusting Problem-Discovery Rates Estimated From Small Samples"

For many years I have promoted the measurement of p rates from usability studies for the dual purpose of (a) projecting required sample sizes and (b) estimating the proportion of discovered problems for a given sample size (Lewis, 1982, 1994, 2001).

I have always made the implicit assumption that values of p estimated from small samples would have properties similar to those of a mean—that the variance would be greater than for studies with larger sample sizes, but in the long run estimates of p would be unbiased. I was really surprised (make that appalled) when Hertzum and Jacobsen (this issue) demonstrated in their article that estimates of p based on small samples are almost always inflated. The consequence of this is that practitioners who use small-sample estimates of p to assess their progress when running a usability study will think they are doing much better than they really are. Practitioners who use small-sample estimates of p to project required sample sizes for a usability study will seriously underestimate the true sample size requirement.

This spurred me to investigate whether there were any procedures that could reliably compensate for the small-sample inflation of p. If not, then it would be important for practitioners to become aware of this limitation and to stop using small-sample estimates of p. If so, then it would be important for practitioners to begin using the appropriate adjustment procedure(s) to ensure accurate assessment of sample size requirements and proportions of discovered problems. Fortunately, techniques based on observations by Hertzum and Jacobsen (this issue) and a discounting method borrowed from statistical language modeling can produce very accurate adjustments of p.

2.6. "Effect of Perceived Hedonic Quality on Product Appealingness"

Within the IBM User-Centered Design community and outside of IBM (e.g., see Tractinsky, Katz, & Ikar, 2000), there has been a growing emphasis over the last few years to extend user-centered design beyond traditional usability issues and to address the total user experience. One factor that has inhibited this activity is the paucity of instruments for assessing nontraditional aspects of users' emotional responses to products. Hassenzahl (this issue) has started a line of research in which he uses semantic differentials to measure both ergonomic and hedonic quality and relates these measurements to the appealingness of a product. Although there is still a lot of work to do to validate these measurements, it is a promising start that should be of interest to practitioners who have an interest in the total user experience.

3. CONCLUSIONS

I hope this special issue will be of interest to both usability scientists and practitioners. The contributors are from both research (Gillan, Hartson, Andre, Williges, and Hertzum) and applied (Bias, Cordes, Jacobsen, Lewis, and Hassenzahl) settings, with two of the articles collaborations between the settings (Gillan & Bias, this issue; Hertzum & Jacobsen, this issue).

The articles in this special issue address many of the topics listed in Section 1.2 (although there is still much work to do). One important outstanding issue, though, is the development of a definition of what constitutes a real usability problem with which a broad base of usability scientists and practitioners can agree. This is a topic that comes up in half of the articles in this issue (Hartson et al.; Hertzum & Jacobsen; Lewis) but is one for which I have not yet seen a truly satisfactory treatment (but see the following for some current work in this area: Cockton & Lavery, 1999; Connell & Hammond, 1999; Hassenzahl, 2000; Lavery, Cockton, & Atkinson, 1997; Lee, 1998; Virzi, Sokolov, & Karis, 1996).

Despite the unanswered questions, I believe that the field of usability engineering is in much better shape than it was 20 years ago (both methodologically and with regard to the respect of product developers for usability practitioners), and I look forward to seeing and participating in the developments that will occur over the next 20 years. I also look forward to seeing the effect (if any) that the articles published in this special issue will have on the course of future usability research and practice.

REFERENCES

Chin, J. P., Diehl, V. A., & Norman, L. K. (1988). Development of an instrument measuring user satisfaction of the human–computer interface. In *Conference Proceedings of Human Factors in Computing Systems CHI '88* (pp. 213–218). Washington, DC: Association for Computing Machinery.

Cockton, G., & Lavery, D. (1999). A framework for usability problem extraction. In *Human–Computer Interaction—INTERACT '99* (pp. 344–352). Amsterdam: IOS Press.

Connell, I. W., & Hammond, N. V. (1999). Comparing usability evaluation principles with heuristics: Problem instances vs. problem types. In *Human–Computer Interaction—INTERACT '99* (pp. 621–629). Amsterdam: IOS Press.

Gray, W. D., & Salzman, M. C. (1998). Damaged merchandise? A review of experiments that compare usability evaluation methods. *Human–Computer Interaction, 13,* 203–261.

Hassenzahl, M. (2000). Prioritizing usability problems: Data-driven and judgement-driven severity estimates. *Behaviour and Information Technology, 19,* 29–42.

Kirakowski, J., & Corbett, M. (1993). SUMI: The Software Usability Measurement Inventory. *British Journal of Educational Technology, 24,* 210–212.

Lavery, D., Cockton, G., & Atkinson, M. P. (1997). Comparison of evaluation methods using structured usability problem reports. *Behaviour and Information Technology, 16,* 246–266.

Lee, W. O. (1998). Analysis of problems found in user testing using an approximate model of user action. In *People and Computers XIII: Proceedings of HCI '98* (pp. 23–35). Sheffield, England: Springer-Verlag.

Lewis, J. R. (1982). Testing small-system customer set-up. In *Proceedings of the Human Factors Society 26th Annual Meeting* (pp. 718–720). Santa Monica, CA: Human Factors Society.

Lewis, J. R. (1994). Sample sizes for usability studies: Additional considerations. *Human Factors, 36,* 368–378.

Lewis, J. R. (1995). IBM computer usability satisfaction questionnaires: Psychometric evaluation and instructions for use. *International Journal of Human–Computer Interaction, 7,* 57–78.

Lewis, J. R. (1996). Reaping the benefits of modern usability evaluation: The Simon story. In A. F. Ozok & G. Salvendy (Eds.), *Advances in applied ergonomics: Proceedings of the 1st International Conference on Applied Ergonomics* (pp. 752–757). Istanbul, Turkey: USA Publishing.

Lewis, J. R. (1999). Trade-offs in the design of the IBM computer usability satisfaction questionnaires. In H. Bullinger & J. Ziegler (Eds.), *Human–computer interaction: Ergonomics and user interfaces—Vol. 1* (pp. 1023–1027). Mahwah, NJ: Lawrence Erlbaum Associates, Inc.

Lewis, J. R. (2001). *Sample size estimation and use of substitute audiences* (Tech. Rep. No. 29.3385). Raleigh, NC: IBM. Available from the author.

Lilienfeld, S. O., Wood, J. M., & Garb, H. N. (2000). The scientific status of projective techniques. *Psychological Science in the Public Interest, 1*, 27–66.

Molich, R., Bevan, N., Curson, I., Butler, S., Kindlund, E., Miller, D., & Kirakowski, J. (1998). Comparative evaluation of usability tests. In *Usability Professionals Association Annual Conference Proceedings* (pp. 189–200). Washington, DC: Usability Professionals Association.

Nielsen, J., & Landauer, T. K. (1993). A mathematical model of the finding of usability problems. In *Conference Proceedings on Human Factors in Computing Systems—CHI '93* (pp. 206–213). New York: Association for Computing Machinery.

Tractinsky, N., Katz, A. S., & Ikar, D. (2000). What is beautiful is usable. *Interacting With Computers, 13*, 127–145.

Virzi, R. A. (1990). Streamlining the design process: Running fewer subjects. In *Proceedings of the Human Factors Society 34th Annual Meeting* (pp. 291–294). Santa Monica, CA: Human Factors Society.

Virzi, R. A. (1992). Refining the test phase of usability evaluation: How many subjects is enough? *Human Factors, 34*, 443–451.

Virzi, R. A., Sokolov, J. L., & Karis, D. (1996). Usability problem identification using both low- and high-fidelity prototypes. In *Proceedings on Human Factors in Computing Systems CHI '96* (pp. 236–243). New York: Association for Computing Machinery.

INTERNATIONAL JOURNAL OF HUMAN–COMPUTER INTERACTION, 13(4), 351–372
Copyright © 2001, Lawrence Erlbaum Associates, Inc.

Usability Science. I: Foundations

Douglas J. Gillan
Psychology Department
New Mexico State University

Randolph G. Bias
Austin Usability

In this article, we describe and analyze the emergence of a scientific discipline, usability science, which bridges basic research in cognition and perception and the design of usable technology. An analogy between usability science and medical science (which bridges basic biological science and medical practice) is discussed, with lessons drawn from the way in which medical practice translates practical problems into basic research and fosters technology transfer from research to technology. The similarities and differences of usability science to selected applied and basic research disciplines—human factors and human–computer interaction (HCI) is also described. The underlying philosophical differences between basic cognitive research and usability science are described as Wundtian structuralism versus Jamesian pragmatism. Finally, issues that usability science is likely to continue to address—presentation of information, user navigation, interaction, learning, and methods—are described with selective reviews of work in graph reading, controlled movement, and method development and validation.

1. INTRODUCTION

Our purpose in this article is to describe and analyze the emergence of a new applied scientific discipline, usability science. We believe that usability science covers the conceptual area between the basic cognitive and behavioral sciences (primarily cognitive and perceptual psychology) and usability engineering. Our examination of this nascent field describes (a) a rationale for its emergence, (b) analogical comparisons with other applied sciences, (c) differences between usability science and related basic and applied fields, (d) the philosophies underlying the relevant basic research fields and usability science, and (e) selected critical issues around which usability science appears to be organizing itself. As with any observational study, the raw observations we make about the emergence of usability science may lend themselves to alternative interpretations. We neither mean to preclude any other in-

Requests for reprints should be sent to Douglas Gillan, Psychology Department, Box 3452, New Mexico State University, University Park, NM 88005. E-mail: gillan@cri.nmsu.edu

terpretations nor to claim that our interpretations are the final word. Rather, we simply hope that the observations and interpretations in this article can stimulate thinking about an important new field of applied science.

1.1. Relation Between Science and Design

The relation between basic science in cognition and perception and the design of usable interfaces has been a topic of some interest in human–computer interaction (HCI; e.g., see Carroll, 1991, 1997; Landauer, 1987; Long, 1996; Newell & Card, 1985; Norman, 1987). These discussions have focused on three major points. First, authors have examined the direct applicability of theories, models, empirical laws, and basic research findings to the design of usable interfaces (e.g., Carroll & Kellogg, 1989; Landauer, 1987). One frequently cited example of an empirical law from cognitive psychology that might apply directly to user interface design is Fitts's law (see MacKenzie, 1992, for a review). Despite claims about its direct applicability (e.g., Card, Moran, & Newell, 1983), the specific application of Fitts's law to predict movement times depends on a detailed analysis of the user's interaction (e.g., Gillan, Holden, Adam, Rudisill, & Magee, 1990, 1992), as we discuss in Section 5 of this article. Possibly because its application is conditional on the details of the user's interaction, it is difficult to identify a design feature that has been directly and uniquely derived from Fitts's law. Unambiguous examples in which designers directly applied other theories, models, laws, or experimental results from basic cognition or perception are likewise elusive. We do not mean to imply that Fitts's law has had no value in HCI: Researchers have used it to compare input devices (e.g., Card, English, & Burr, 1978). Thus, basic research may be useful in explaining why certain interface features are superior to others in usability, but such research seems to be used rarely, if ever, in the development of those features.

The second major focal point for the interplay between basic research in experimental psychology and user interface design and evaluation has been the application of methodologies. Bailey (1993) found that user interface designers with training in cognition created more usable designs than did those without such training. One element of training in cognition that may have helped produce better designers is the sets of skills acquired through creating and running experiments, as well as analyzing data. Although cognitive experiments differ markedly from usability design and evaluation, they also share common features of data collection, analysis, and interpretation. Creating novel artifacts involves both (a) a generative step in which the designer produces new ideas and approaches and (b) an evaluative step in which the designer discriminates the good from the bad ideas (e.g., Weber, 1992). Training in experimental methods may improve the skills and conceptual thinking required for the evaluation of usable artifacts.

The third major focus in discussions of the relation between psychological research and interface design concerns the flow of information from design and evaluation to basic psychological theory and research. For example, Landauer (1987) and Gillan and Schvaneveldt (1999) proposed that problems that users have interacting with systems could be used to inform cognitive theories and research. More formal proposals along these lines come from Carroll and Kellogg (1989) and Woods (1998). These authors suggested that we can consider each design to be the designer's implicit theory about aspects of human perception and cognition. Thus, one purpose of psychological science should be to unpack and compare the theoretical claims that underlie the design of usable and unusable interfaces (e.g., Carroll & Campbell, 1989).

As a consequence of the limited instances of direct application of cognitive and perceptual theories, models, and experiments to the design of usable computing systems, we hear frequent complaints from usability engineers that they cannot use basic research from cognitive and perceptual psychology (see Bias & Gillan, 1998; Gillan, 1998). The absence of a substantial impact of cognition and perception on interface design can be seen as a failure of technology transfer. In previous articles (Gillan & Bias, 1992; Gillan & Schvaneveldt, 1999), we examined several reasons for this failure, including differences between research and design in mission, timeline, rewards, activities, and modes and types of communication.

1.2. Roswell Test and Applying Cognitive Psychology

The serious failure of the translation of knowledge from cognitive and perceptual research to HCI design or to any other real-world application can be seen by conducting a thought experiment, which we call the *Roswell Test*. Imagine that aliens from outer space had crashed their UFO at Roswell, New Mexico in 1948 and that they were only part of a larger invasion force. Furthermore, imagine that the invasion force had decided to conquer the world by eliminating progress in one science during the next 50 years. Now, think about how your everyday life might have been changed if that one science in which no progress occurred had been cognitive psychology. Would your home be different? Your car? Workplace? Schools? Recreational activities? Health care? In contrast, consider how your life might have been changed if there had been no progress in genetics or geology or biochemistry. If cognitive psychology had been on the right track for the last 50 years, one might think that there would be some obvious application that had an impact on our everyday lives. The pragmatic approach to psychology (e.g., James, 1907), which we discuss in more detail later, suggests that the real world often provides the best tests for scientific theories. If that is true, cognitive psychology appears not to have passed many of the tests.

The failure of transfer of knowledge from basic cognitive research also has potentially serious consequences for usability engineering. In the absence of this transfer, designers tend to create designs based on their own assumptions about how people perceive and think rather than based on a set of empirically validated principles. For example, take the ubiquitous use of movement and flashing in advertising on Web sites. Both the movement and light flashes stimulate motion detectors in the peripheral visual system, and as a consequence, automatically attract focal attention to that motion or flash. This automatic shift of attention may be valuable to the advertiser the first one or two times that it occurs. However, the automatic attraction of attention may continue more or less unabated as the motion continues, which users are likely to find annoying. As a consequence, users may leave the site for one that is more congenial to completing their task, so the Web site loses customers (we write this from personal experience—Douglas Gillan refuses to return to certain Web sites containing ads that use excessive motion). Alternatively, users may develop "banner blindness"—that is, they may learn to avoid fixating "their eyes on anything that looks like a banner ad due to their shape or position on the page" (Nielsen, 1999, para 30). Nielsen (1979) suggested that the continuing reduction in click-through rates at Web sites may be due to users developing banner blindness to ads that interrupt their main task. Forcing users to develop an attentional strategy that leads to ignoring information is also not the goal of Web site or ad designers. Rather, an understanding of the psychological principles used directly in the design—say, ads that flashed or moved only once or twice after a user entered a Web site then stopped and statically maintained their spatial location—would lead to more useful Web sites.

In addition to identifying reasons for the failure of translating cognitive and perceptual principles and research into interface designs, we (Gillan & Bias, 1992; Gillan & Schvaneveldt, 1999) have also previously described some approaches to increase this transfer, including instantiating psychological principles as design guidelines and design analysis tools and the use of gatekeeping tools, which might translate cognitive and perceptual research findings into interface design ideas. In this article, we attempt to provide a framework for approaches to technology transfer between experimental psychology and interface design by examining the development of an applied science that mediates between the two disciplines—usability science.

2. ANALOGIES BETWEEN USABILITY SCIENCE AND OTHER APPLIED SCIENCES

Examining a science that has been shown to transfer knowledge successfully between the basic research world and fields of practice provides a point at which to begin a discussion of the relation of basic perceptual and cognitive psychology with usability science. Medical science serves as a successful interface between basic biological science and medical practice, with many of its researchers having dual training as scientists and physicians. In fact, the history of the biological and medical sciences suggests that the distinction between researchers and practitioners was simply not made initially. For example, Ignaz Semmelweis was a practicing physician who, in the 1840s, observed that the rates of death from childbed fever (also known as puerperal fever) were much higher in a maternity ward in which surgeons delivered babies than in a ward in which midwives delivered babies (see Hempel, 1966, for a review). Following the death from childbed fever of a surgeon friend who had cut himself during the autopsy of a childbed fever victim, Semmelweis proposed that an invisible agent caused childbed fever. He instituted policies that required surgeons to disinfect their hands and change their coats after surgery, resulting in a dramatic decline in deaths in the surgeons' maternity ward. (Unfortunately, following a surgeon's complaints about the new policies, Semmelweis was fired. He ultimately died in an insane asylum, the victim of a blood infection that may have been childbed fever.) With the advent of germ theory, biologists came to recognize that the invisible agent that Semmelweis proposed was a *germ*.

Some of the most compelling evidence for germ theory came from Robert Koch, another practicing physician who functioned as a basic research scientist (see Brock, 1999, for a review). Koch was a country doctor who, as he studied the blood of anthrax victims (both cows and humans) in the 1870s, observed a large microbe that he hypothesized might be the causal agent for anthrax. Fearing that, as a rural physician, the established scientists would not accept his hypothesis, he extracted and purified the microbe and then gave it to animals that became infected. He repeated this cycle of isolating the target agent, extracting the agent, using the agent to infect a healthy participant, then isolating the agent from the new victim, ultimately identifying the microbe as the cause of anthrax. Koch's method became the necessary test for establishing that a target agent is the cause of a disease.

Interestingly, not all of the early advances in biology came from medical needs. In the 17th century, van Leeuwenhoek was a Dutch cloth merchant who used the newly invented microscope to examine the weave of the cloth that he was considering buying. Dissatisfied with the quality of the images that he could see from the microscopes, which he bought from others, he began to grind his own microscope lenses. Eventually he looked at a sample of water through one of his lenses and observed microorganisms such as algae and large bacteria moving. Because of their movement, van Leeuwenhoek described these as living creatures. His drawings and descriptions of these microorganisms captivated biologists.

The field of biology today has greater differentiation between basic science and medical practice than in the time of Semmelweis and Koch. However, there remains a strong flow of ideas and information between applied problems and basic research, with medical scientists serving as the bridge. For example, Stanley Prusiner initiated his Nobel Prize-winning basic research on a protein-based infectious agent (called a *prion*) in 1972 when one of his patients died from Creutzfeldt–Jacob disease (CJD; see Rhodes, 1997). Prusiner provided evidence that a folded form of a prion may cause a wide range of neural diseases from scrapie in sheep to bovine spongiform encephalopathy (widely known as mad cow disease) to kuru among the indigenous people of Papua, New Guinea, to both the traditional and variant forms of CJD. Although the research on prions has not yet yielded any medical therapies or inoculations, preventative measures (e.g., bans on religious practices in New Guinea and on British beef in Europe) have followed from the research.

Medical science and its relations to basic research in biology and medical practice serves as an imprecise analogy for usability science. As the history of medicine and biology reviewed previously suggests, medical science, biology, and medical practice all evolved from a common ancestor, therefore their linkage is part of their lineage. In contrast, usability science is in the process of evolving now, over a century after basic research in psychology, more than 30 years after cognitive psychology gained preeminence in experimental psychology (e.g., Neisser, 1967), and several decades after usability engineers began to practice. In addition, the biological and psychological sciences differ in numerous ways, including the relations between measures and constructs, the variability of dependent variables, the physical control over both independent and extraneous variables, the ability for convergent validation, and the reliability of phenomena.

Despite the imprecision of the analogy between the biological and cognitive sciences, we believe that there is heuristic value in examining the relations among basic biological science, the more applied medical science, and the completely applied medical practice—specifically, that such an examination can reveal guidelines for applying science. (Note that a more detailed examination focused on a wide variety of applied sciences, including medical science, might extend these insights.) The first guideline from the review of the application of biology to medical practice is that interesting basic research problems can be identified from knowing about real-world problems. Donald Broadbent, a founder of cognitive psychology, also recognized this.

> "Briefly, I do not believe that one should start with a model of man and then investigate those areas in which the model predicts particular results. I believe one should start from practical problems, which at any one time will point us toward some part of human life" (Broadbent, 1980, pp. 117–118).

A second guideline suggests that observational methods can be critical in applied science. In fact, good observational technique may be as important as training in the specific scientific domain, as was true of van Leeuwenhoek. Third, in counterpoint to the second observation, at least some practitioners need to be able to understand and apply the scientific method so they can make causal inferences about processes and mechanisms. This guideline for applying science was especially important in the case of Koch's identification of the cause of anthrax. Fourth, general methodological approaches and conceptual or mechanistic ideas can organize and guide applied science. As a consequence, applied science can serve as a driving force for progress in a discipline; the applied science need not be merely reactive to events in the related basic science or applied practice.

Figure 1 provides a simple graphical representation for the interrelations among basic science, applied science, and field of practice. In addition to the four general

FIGURE 1   A conceptual model of the relations between basic science, applied science, and fields of practice. The model shows that some practical issues pass through applied science and are translated into basic research problems (indicated by the solid arrows between fields and the stippled arrow within applied science); some basic research is translated into practical applications (indicated by the solid arrows between fields and the stippled arrow within applied science); and some problems go from practice to applied science without being passed on to basic research (indicated by the striped arrows).

principles proposed previously, Figure 1 indicates an important fifth principle of applied science: Not all issues that go from practice to science end up as basic research problems. An example from medical practice might be the development and evaluation of surgical procedures. Surgeons might develop a new procedure (e.g., for heart bypass operations), then applied researchers would evaluate the efficacy and safety of the procedure relative to existing practice. Typically, in cases such as this, no broad basic issues would be addressed, so only applied researchers would become involved. As we discuss later, evaluating usability engineering methods may be a parallel case, with usability scientists but not basic cognitive psychologists taking the primary interest.

3. RELATION BETWEEN USABILITY SCIENCE AND OTHER DISCIPLINES

The previous discussion begins to distinguish the function of usability science from the functions of both basic cognitive psychology and usability engineering. Cognitive psychology focuses on describing general principles and understanding the mechanisms of high-level perception, memory, thinking, language, and so forth. On the other hand, usability engineering centers around the design, development, and evaluation of usable technological artifacts. Usability science can serve, at least in part, as the means by which (a) problems identified by usability engineers can be translated into more basic research issues, and (b) theories, empirical laws, and empirical findings can be put into terms usable for design, development, and evaluation. Accordingly, usability science needs to be more general than usability engineering, working on overarching issues that should influence many different specific artifacts. (We describe some of those overarching issues later in Section 5.) In contrast, usability science needs to be more specific than cognitive psychology and cognitive science, focusing on making technological applications easier to use.

What about other disciplines that seem to fall in the middle ground between psychology and technology, specifically human factors and HCI? Actually, previous and ongoing research that might be considered to be in the domain of usability science has been done by researchers in human factors and HCI. However, we believe that the emerging usability science as a discipline differs from these others. We start by asserting that if human factors and HCI had been an effective bridge between basic research in cognitive psychology and the developers of usable technology, there would not be so much technology that is so difficult to use. Also, usability engineers might not report that they do not typically make use of information from basic cognitive psychology (e.g., Gillan, 1998). Yet usability science needs to be more than simply a recasting of these two older disciplines. The set of issues that we believe define usability science should differ in content and emphasis from those that have defined human factors and HCI.

Examining these disciplines may help to differentiate them from usability science. Human factors is a broad field concerned with "the study of how humans accomplish work-related tasks in the context of human–machine system operation, and how behavioral and nonbehavioral variables affect that accomplishment" (Meister, 1989, p. 2). Human factors deals with many issues that are outside of the purview of usability science (at least as we see it), including anthropometrics and biomechanics. In addition, the focus of human factors on work-related tasks and on machines is narrower than in usability science that would not be restricted to work and might study nonmechanical artifacts that humans use, such as a university campus. For example, many Web sites involve extracting information relevant to the users' personal lives rather than their work, but those users should find those Web sites easy to use.

Consider that usability science is merely an extension of human factors outside of the realm of working with machines, and consider the important difference of the training of human factors professionals, as it is currently practiced, versus our vision of training of usability scientists. Almost all human factors professionals who receive explicit training at a university are in engineering departments (most frequently, industrial engineering) or psychology departments—of the 87 graduate programs listed in the Human Factors Directory for Graduate Programs (Human Factors and Ergonomics Society, 2000), 37% (i.e., 32 programs) are in psychology departments and 48% (i.e., 41 programs) are in engineering programs. One problem with these two approaches to human factors is that the engineering programs, in general, train students to apply rules or algorithms as a way to implement science, with minimal training in research. In contrast, psychology departments tend to teach research methods and basic principles in experimental psychology but much less about usability design and evaluation, compromising the psychology students' ability to bridge basic research and practice. As a bridge between basic psychological research and usability practice, we envision a cadre of usability scientists trained to both conduct research and design usable artifacts located in various universities, companies, and government agencies.

The field of HCI differs from usability science because of the former's exclusive focus on computing and often on the interactive aspects of using computers to the exclusion of less interactive aspects such as reading displays. In contrast, usability science considers the usability of both computing and noncomputing technologies. Norman's (1988) description of doors that people cannot figure out whether to push or pull provides an example of such difficult to use, low-tech artifacts. In addition, the field of HCI in recent years has been dominated to an extent by the development of technologies that could possibly lead to increased usability but that have not been demonstrated to do so. Gillan and Schvaneveldt (1999) distinguished between need-based technological development (i.e., technology based on the needs of the users) and designer-based development (i.e., technology based on the technological capabilities of the designers). HCI has focused more on designer-based than on need-based development. For example, we examined all of the articles from two randomly selected recent computer–human interaction (CHI) conferences—1995 (Katz, Mack, Marks, Rosson, & Nielsen, 1995) and 1998 (Karat, Lund, Coutaz, & Karat, 1998)—and placed each article into one of the following six categories: (a) description of a need-based HCI development project, (b) user-focused articles (including user performance and user models), (c) user evaluation studies (including competitive evaluations, evaluation methods, and evaluation case studies), (d) description of a technology-based HCI development project with no user data at all, (e) description of a technology-based HCI development project with user test data (this ranged from 2 users providing qualitative responses to a full-scale user test), and (f) descriptions of design method and case studies. (We categorized articles, but excluded posters, short reports, and design briefings.) Table 1, which shows the categorization from the 2 years of CHI, reveals the same basic pattern for both years—virtually no articles described development projects that had a clear empirical basis in the needs of users, but well over half of the articles described technology-based development projects. Within those technology-based development projects, about half had no user data and half had at least some post hoc user support for the development. If we consider the first three categories to be user focused and the latter three categories to be technology focused, then across the 2 years 48 articles were user focused and 101 articles were technology focused. Based on the binomial distribution, with the assumption that half of the articles should be of each type, this split between the two types would happen by chance less than 5% of the time. In other words, the CHI conference proceedings articles support the previous assertion that HCI, at present, has a strong technology focus. This is in contrast to the user focus we propose for usability science.

4. PHILOSOPHICAL UNDERPINNINGS

During the past 20 years, the issues that have dominated theory and research in basic cognitive and perceptual psychology have focused on identifying the elements of the mind and their relations. These issues include the architecture of cognition (e.g., Anderson, 1983, 1993; McClelland & Rumelhart, 1986), the number and characteristics of different memory systems (e.g., Tulving, 1985), the mental modules in language and other cognitive processes (Fodor, 1983), and the neuropsychological correlates of cognition (e.g., Squire, Knowlton, & Musen, 1993).

The focus of modern cognitive psychology on the elements of the mind resembles structuralism (Gillan & Schvaneveldt, 1999; Wilcox, 1992), a philosophical approach to psychology developed in Germany late in the 19th century by Wundt and popularized in the United States by Titchener (1898) in the early 20th century. Wundt and Titchener appear to have been influenced by physical chemistry in proposing that consciousness could be decomposed into mental elements and that the role of experimental psychology was to catalog those elements and

**Table 1: Number of Articles Concerned With Various Topics
in the CHI '95 and CHI '98 Conference Proceedings**

Categories of Articles	1995	1998
User need-based development	0	1
User performance and models	16	23
User evaluation	7	1
Technology-based development (no user data)	16	23
Technology-based development (with user data)	23	23
Design method and experience	6	10

Note. CHI = computer–human interaction.

to understand the ways in which they combined to create mental structures (e.g., Titchener, 1898). In addition, the structuralists of that time believed that psychological science should not attempt to be applied, but should seek knowledge for its own sake (Neel, 1969). Note that a popular movement in philosophy, anthropology, and literary criticism during the 1960s and 1970s was also called *structuralism;* however, this version of structuralism came from Saussure's approach to linguistics that focused on the role of structure in the analysis of language (e.g., Gardner, 1972). We believe that the cognitive psychologists of today have adopted the Wundtian version of structuralism. One of the criticisms of Wundt–Titchener structuralism was that it was too passive and inward focused, with no concern for the cognition of real people as they interacted with the real world. We believe that a similar criticism can be directed toward the neostructuralism of today's cognitive research.

In the United States, opposition to structuralism in the late-18th and early-19th centuries coalesced around two related philosophical approaches—functionalism and pragmatism. Dewey (1896) proposed that the recently described reflex arc could serve as a metaphor for the basic unit of the mind; this idea provided the impetus for the development of functionalism. Thus, in contrast to structuralism, functionalism proposed that the purpose of the mind was to transform perceptual inputs into behavioral consequences. As functionalism developed, it came to focus on how organisms used these stimulus–response acts to mediate between the world and their goals or needs (e.g., Angell, 1907). Functionalism focused on the functions of thought, with practical applications being one of the important aspects of the development and testing of theories (Carr, 1925).

Charles Peirce (1878) developed the philosophical doctrine of pragmatism as a theory that proposed that how people think is a function of what they are prepared to do. Peirce proposed that thought is valuable only to the extent that it produces observable and generally agreed-on useful actions. William James (1907) extended Peirce's doctrine (and, according to Peirce, ruined it) by suggesting that thought is valuable if it produces an outcome in the world that is of value to the actor. James (1975) suggested the question "What special difference would come into the world if this idea were true or false?" (p. 268) lay at the heart of pragmatism. In other words, James's (1907) version of pragmatism suggested that the mind functions pragmatically, with people selecting to think in ways that can lead to personally valuable outcomes.

A second sense of James's (1975) question is often taken as the single hallmark of pragmatism (e.g., Sternberg, 1999): What special difference would come into the world if a scientific idea were true or false? According to this view, the way to evaluate a scientific idea is based on its practical value in the world. As we suggested in the discussion of the Roswell Test, a related way to think about this is that the world provides the ultimate testing grounds for scientific theories.

Clearly, the second sense of pragmatism, that science must be applicable to be useful (or, as we claimed earlier, that the value of a theory is proved by its application in the world), is a central tenet of usability science. The research of usability scientists will have value to the extent that it affects the creation of usable technology. In this principle alone, usability science differs from basic research in cognitive science. However, we also believe that the sense that James (1907) originally intended for his question—that pragmatic concerns guide our thinking—is an important principle of usability science. Accordingly, usability science should be considered to be a pragmatist science.

If usability science takes seriously the notion of pragmatism as the basis of cognition, how would that inform the science that we do? We believe that it reorients our model of cognition in the following four ways: (a) cognition is goal oriented; (b) cognition helps us adapt to the world; (c) cognition is closely tied to actions and their consequences in the world; and (d) the physical, social, cultural, and task contexts of the world in which cognition occurs set the conditions for cognition. Conducting research that removes goals, adaptation to the world, action and consequences, and the various kinds of context as key parameters leads to research that may be only generalizable to a narrow set of circumstances defined by that experiment. Yet, the typical experiment in cognitive psychology during the past 20 years has controlled all of these features as potential contaminants in the investigation of the basic cognitive elements. The desire to control these features in a research environment can lead to research that only describes cognition for those narrow circumstances, with little possibility to make contact with the cognition of real people acting in a real or a virtual world. In the next section, we point out how research that defines usability science has attempted to address, rather than control, some of these experimental foci of pragmatism.

5. SELECTED ISSUES FOR USABILITY SCIENCE

In this section, we propose that certain issues in usability science will be enduring ones. However, our ability to gaze into the future will undoubtedly be imprecise; therefore, we do not claim that the issues described here will (or should) circumscribe the emerging discipline of usability science. Undoubtedly, based on the model shown in Figure 1, issues will arise from users' needs, from the needs of the usability design and development community, and from basic research in perception and cognition. However, due to the nature of human perception and cognition, as well as the ways in which people interact with technology, certain general issues are likely to be important on a continuing basis for designing the usability of any system. In addition, usability science will be likely to make major contributions to our understanding of these issues, also on a continuing basis. These base issues in usability science include (a) the presentation of information by a system to the user (covering such topics as the effects of the user's task on how to present information, the role of user expertise, and the perceptual capabilities and disabilities of the user), (b) user navigation through physical and informational spaces (including spatial metaphors for navigation and models for information retrieval), (c) modes and methods of interaction with a system (with specific topics such as the effects of automation and the consequent active versus passive control, attention in interacting with the system, command language vs. direct manipulation, and motor control), (d) acquiring knowledge and transferring knowledge of use from one system to others (including user mental models of systems and tasks, instructional methods, active learning of a system, minimalism in training, and designing for errors), and (e) the development and adaptation of methods (including identifying methods that could be helpful to usability engineers and validating the application of those methods). A thorough review of any one of these content areas of usability science exceeds the scope of this article—each area deserves a separate review article. Accordingly, the following sections provide a very selective review of research in three areas, with two sources of bias motivating the selections: The research reflects content areas of interest to the authors and exemplifies the pragmatist approach to science.

5.1. Presentation of Information

In recent years, one major area of research on presentation of information has been the graphical display of quantitative information. Much of the research on this topic has involved comparisons of types of displays to determine which produces the best performance. However, over the past 10 to 15 years, some researchers have focused on understanding the perceptual and cognitive processes involved in graph reading. This approach to graph reading is of interest here in part because it makes use of a neostructuralist approach to meet a pragmatist agenda. Specifically, a finding related to pragmatism from this graph reading research is that the cognitive and perceptual processes used to read graphs vary as a function of many different contextual variables: the task, the specific features of the graph, and the graph reader's knowledge, strategies, and culture.

The task context influences graph reading in several ways. When people use a specific graph (e.g., a line graph), they do not apply the same perceptual and cognitive processes across all tasks. For example, Gillan and Lewis (1994) and Gillan and Neary (1992) found that reading a line graph to compare two points involved visual search and spatial comparison, whereas reading a line graph to determine the difference in value between the same two points involved visual search, encoding of values, and arithmetic computation. Similar task-dependent differences in application of processing components have been shown with bar graphs (e.g., Gillan & Neary, 1992; Lohse, 1993) and pie graphs (Gillan, 2000a; Gillan & Callahan, 2000; Simkin & Hastie, 1987). The task context may also interact with the features of the graph. Carswell and Wickens (1988; see also Carswell, 1992; Wickens & Carswell, 1995) found that focused tasks (e.g., tasks that involved processing only single data points or simple two-indicator comparisons) were most effectively performed if the graph was a separable graph, such as a bar graph. In contrast, a graph reader's performance in an integral task (e.g., tasks involving global comparisons or synthesizing across multiple data points) was better if the graph was an object-like integral graph, such as a line graph, rather than a separable graph.

The usefulness of object-like graphs to perform certain tasks turns out to be influenced by the perceptual strategies of the reader, as well as by the task. In research on star graphs (multivariate graphs consisting of lines radiating from a central point, with the length of each line indicating an amount), Gillan and Harrison (1999) observed graph readers using two different types of perceptual strategies to identify the different star in a set of four (the difference was that one of five lines that made up the star was slightly longer in one of the stars). One type of reader showed no difference in the time to identify the different star as a function of the position of the longer line on the different star; in contrast, other graph readers showed a pattern of taking longer to decide which star was different as the longer line was further off vertical. These two types of graph readers correspond to Cooper's (1979) Type I (holistic) and Type II (analytic) perceivers, respectively. Of most interest, adding object-like features (an outline of the star made by joining the ends of each line segment) decreased the time to identify the different star in a set of four only for the holistic perceivers. Thus, the context of a graph reader's perceptual strategy also influences the perceptual and cognitive processes that the reader applies.

The graph readers' knowledge is another critical contextual variable that needs to be taken into account in trying to understand both the processes of graph reading and how to design graphs. Gillan (1995) trained people in a perceptual method for estimating the mean of a set of data points represented in a bar or line graph by determining the graphical midpoint of the relevant data and then finding the value of the midpoint. This perceptual method was labeled *visual arithmetic*. In contrast, Gillan and Lewis (1994) found that people without training in visual arithmetic would estimate a mean by searching for each indicator, encoding each indicator's value from the y-axis, then performing the mental arithmetic on the set of values necessary to find the mean. Gillan (1995) showed that graph readers trained in visual arithmetic were both faster and more accurate to estimate a mean from a graph than were those who received no training but that the improved performance did not generalize to other tasks. Thus, graph readers procedural knowledge will influence the component processes that they apply to complete a given task with a graph.

Finally, the cultural context of the graph readers can influence how people read graphs. Tversky, Kugelmass, and Winter (1991) found that graph readers from cultures with a printed language that is read left to right had better comprehension of a graph that had the x-axis ordered with the lowest value at the far left of the axis and ascending values rightward; in contrast, graph readers from cultures with a printed language that is read right-to-left had better comprehension of a graph with an x-axis organized in the right-to-left. All graph readers had better comprehension when the y-axis was ordered with low values on the bottom and high values on the top, independent of their culture.

These experimental studies in the processes of graph reading have led to an enhanced understanding of both the perceptual and cognitive processes that underlie graph reading performance and how various kinds of contextual influences modulate the reader's application of those processes. In addition, these studies—in line with the second sense of pragmatism described previously—have led to specific applications intended to result in improved graph production. Lohse's (1993) studies and model served to guide the development of a computer-based tool that analyzes graphs and makes recommendations for improvements based on performance predicted by the model. The research described previously by Gillan, Carswell, and Wickens (Gillan, Wickens, Hollands,

& Carswell, 1998) served as an important basis for a set of guidelines for graph design for publications of the Human Factors and Ergonomics Society. Finally, Gillan and Callahan (2000) used their model and research on pie graphs to develop new versions of the pie graph designed to eliminate the need for graph readers to apply certain processing components.

5.2. Interaction and Control

Our review of this topic is focused on the importance of understanding the user's goal, even in very basic interactions with technology. One of the simplest and most heavily studied acts in HCI is using a mouse to move a cursor in a straight line to a target. The time required to perform this act has been described using Fitts's law: $MT = a + b[\log_2(\text{distance}/\text{target size})]$ (see Card et al., 1978, for an early application of Fitts's law to HCI, and MacKenzie, 1992, for a review). However, even for this simple act, knowing the user's goal is essential.

One common application of Fitts's law to HCI has been pointing in text editing (e.g., Card et al., 1978, 1983). In text editing tasks, users point at a block of text with the goal of selecting that text. As Card et al. (1978) modeled this, participants pointed at a text block of varying size, with the size of the block of text serving as the target size in Fitts's law equation. However, Gillan et al. (1990, 1992) pointed out that most word processing programs use a point-and-drag selection method and that the user's goal is not to point at an entire block of text but to point at the initial letter of a block of text across which he or she will drag the cursor. As a consequence of the point-drag sequence, the size of the user's target (the initial letter) does not change size as the text block changes size. In a series of experiments, Gillan et al. (1990, 1992) showed that pointing time was sensitive to changes in the size of the text block when the user's goal was to select the entire text block by clicking on it (as in a point-click sequence), but was not sensitive to the size of the text block when the goal was to stop pointing at the initial letter and to select the text by dragging (as in a point–drag sequence).

Of interest from the Gillan et al. (1990, 1992) experiments was the observation that dragging was slower than pointing, and the slope parameter estimate (b) was markedly higher for dragging than for pointing—355 versus 182, respectively (see also Inkpen, Booth, & Klawe, 1996). The higher slope parameter estimate for dragging than for pointing suggests that, for dragging, movement time is especially sensitive to changes in the index of difficulty for the movement.

Dragging and pointing differ in several different ways. For example, dragging occurs with a finger holding down a mouse button, whereas pointing occurs without a finger holding down a button. However, research by MacKenzie, Sellen, and Buxton (1991) indicates that this variable accounts for only a small part of the difference between the two movements. In their experiment, the only difference between dragging and pointing was the finger position; although they found a significant difference between the Fitts's law slope parameter estimate, that difference (249 for dragging vs. 223 for pointing) was much smaller than that in the Gillan et al. (1990, 1992) research. Thus, we might consider that about 15% of the difference between pointing and dragging is due to the difference in the physical movement. What might account for the remaining difference? One possible explanation concerns differences in the users' goals when they point and when they drag. An error in pointing in normal word processing typically leads to only a small problem—moving the cursor from the end of the pointing movement to the target. In contrast, an error in dragging in normal word processing leads to a much greater negative consequence—restarting the drag movement from the initial point. Thus, one might expect that users would have different goals in the two movements due to the different consequences of erroneous movements—not being too worried about missing a pointed-at target but being more concerned about missing a dragged-to target. A greater focus on accuracy as a goal in dragging than in pointing would likely lead to a different trade-off between speed and accuracy for the two movement types, resulting in different fits to Fitts's law.

Another, very different study provides evidence of the role of the user's goal in understanding how to fit Fitts's law to HCI. Walker and Smelcer (1990) compared selection times between pull-down menus and walking menus. Their analysis of these menus examined two interesting design features: (a) increasing the size of targets as a function of their distance from the cursor, thereby reducing the Fitts's law index of difficulty and, consequently, the movement time; and (b) creating impenetrable borders that backed the menus such that a user could not move the cursor beyond the menu border. They found that both features decreased movement times. Perhaps the most interesting aspect of the research was that the impenetrable borders eliminated the logarithmic relation between movement time and the index of difficulty because users could reach the target with a single ballistic movement. In other words, the impenetrable boarders permitted users to change their goal from trading off speed and accuracy to focusing solely on speed—the design had eliminated the need to be concerned about accuracy.

5.3. Methodology

As we suggested in the discussion of medical science, one persistent issue for any applied science concerns the development and evaluation of methods. Gillan (2000b) has proposed that usability methods can be organized according to three dimensions: (a) the original genesis of the method (e.g., from basic cognitive psychology, basic anthropology, industrial engineering, or usability engineering practice), (b) the function of the method (data collection, data description, data organization, or data generalization), and (c) the point in the design cycle where the method is used (e.g., predesign requirements analysis, design evaluation, or system testing). Researchers and methodologists may find it useful as they develop or adapt methods to consider them within this multidimensional space.

The use of a spatial metaphor for organizing an analysis of usability methods provides an advantage in thinking about the development of new methods. One way in which new methods might be considered is through the addition of a new dimension in the space. For example, most methods have been developed for collecting and analyzing data from an individual user. Thus, the multidimensional space for methods does not have a dimension concerned with the number of users. However, teamwork is an increasingly important topic—from computer-supported cooperative work on the technological side (see Olson & Olson, 1999, for a review) to team tasks, team cognition, and team situational awareness on the users' side (e.g., Cooke, Stout, Rivera, & Salas, 1998). Accordingly, the three-dimensional usability methods space may expand as usability methods expand to incorporate evaluating team knowledge or preferences.

A second way in which a multidimensional methods space might be useful for methodological development would be to use it to identify opportunities for new methods. Usability scientists could examine the methods space for areas that are currently empty. Researchers could then consider the potential uses for a method from that area and, if such a method seemed to provide novel and valuable capabilities for data collection, organization, or analysis, then the method might be developed.

As new methods are brought into the usability tool kit, either because they are incorporated from other fields or they are developed within usability science or engineering, one of the most important functions of usability science will be to determine their validity (Gillan, 2000b). The topic of validation of methods has already been an important one for usability science. For example, researchers have begun to use multivariate statistical techniques, such as the Pathfinder network algorithm, to organize and analyze sequences of user behavior (Cooke, Neville, & Rowe, 1996; Cooke & Gillan, 1999; Gillan & Cooke, 1998). Having tools that provide visual representations of units of user behavior can be helpful to interface designers, for example, to arrange controls spatially that will be used close together in time. In addition, being able to organize sequences of behavior into meaningful units may provide insights into the procedural knowledge of users (see Cooke et al., 1996). However, the successful use of multivariate techniques such as Pathfinder requires research validating that the output of the statistical analysis (in the case of Pathfinder, a network representation of the behavior) is related to other measures of performance (e.g., time to complete the task, number of procedural steps), a form of validation called *convergent validity* (see Gillan & Cooke, 2000, for a discussion of convergent validity).

Gray and Salzman (1998) recently conducted a detailed examination of the validity of selected usability methods that are already in use. They reviewed and evaluated the claims of five studies concerning validity (Desuvire, Kondziela, & Atwood, 1992; Jeffries, Miller, Wharton, & Uyeda, 1991; Karat, Campbell, & Fiegel, 1992; Nielsen, 1992; Nielsen & Phillips, 1993) that compared such usability methods as heuristic evaluations; cognitive walk-throughs; individual walk-throughs; team walk-throughs; user testing; and goals, operators, methods, selection modeling. Gray and Salzman (1998) evaluated these studies designed to compare usability methods for four threats to validity taken from Cook and Campbell (1979): (a) statistical conclusion validity, which is the ways in which researchers use statistics to decide that a manipulated or predictor variable (e.g., type of test) is related to the outcome variable (e.g., number of usability problems identified); (b) internal validity, which is the ability of the research design and statistical methods to allow the researcher to conclude that the manipulated variable caused observed changes in the outcome variable; (c) construct validity, which is the relations of the predictor variable and the outcome variable with real-world variables of interest; and (d) external validity, which is the ability to generalize from the results of specific research to various targets (population of users, types of usability tests, measures, and sets of control features). In addition to the four threats to validity taken from Cook and Campbell (1979), Gray and Salzman (1998) reviewed the studies with regard to the threat to conclusion validity. This threat concerns research reports that make claims that either are not addressed by the research or are contradicted by the research. The five research reports that Gray and Salzman reviewed tended to provide advice to readers who might want to apply the results of their research, but the reports often failed to distinguish between the advice and conclusions that were directly supported by the research and those that were not.

Several comments regarding the issue of threats to validity in the five studies reviewed by Gray and Salzman (1998) seem appropriate here. Validation of a method should be viewed as a process, not a single experiment. Any individual experiment or research study will be flawed, with many of the flaws coming from the inevitable trade-offs necessary to implement the original research question or hypothesis. Evaluating any individual study designed to compare or validate usability methods requires an examination of major violations concerning the threats to validity (e.g., a confound that made it impossible to know if the differences in the number of usability problems observed were caused by the type of test). However, evaluating a study also requires sensitivity to the seriousness of a violation. Ultimately, the key question is whether the value of the information derived from the study outweighs whatever flaws the researchers made in conducting the study. Usability scientists will need to help usability engineers sort through all of the information to come to decisions about the quality of the testing instruments they can select.

5.4. Summary of Issues

In Section 5, we provided a brief overview of selected research in three content areas of usability science: (a) presentation of information, (b) interaction and control, and (c) methodology. Given more space and an unlimited attention span in our readers, we could have made similar arguments for the pragmatist approach in the areas of user navigation and the modes and methods of interaction with a system. In addition, readers will likely generate their own candidates for critical issues in usability science; other issues will be aborning as new technologies and the ever changing culture lead to changes in users' needs or as breakthroughs in basic research modify our understanding of perception and cognition. Also, as these areas grow in number and breadth, it will fall to usability scientists to facilitate accurate and efficient bridges between the research and artifacts.

6. CONCLUDING REMARKS

Will usability science continue to emerge and eventually take its position between basic research and the design of usable technology? We believe that the success of usability engineering will ultimately depend on the development of a systematic and broadly applicable body of knowledge that relates principles of human cognition and perception to the design, development, and evaluation of usable techno-logical artifacts. The emergence of usability science creates two interfaces, just as any bridge does: between cognitive science and usability science at one end and be-tween usability science and product design at the other end. Usability scientists will need to continue taking on the task of extending the basic research from one end of the bridge to the other, interpreting the research for engineers, and transforming the needs of users, as identified by usability engineers, into workable scientific hypoth-eses that can be examined either by basic researchers in cognition or by usability sci-entists themselves.

What can those who believe in the importance of usability science do to enhance the likelihood that it will flourish? First, individuals' behavior and beliefs need to change; those who are doing work on usability science need to communicate the im-portance to usability engineers of having a strong and applicable science base and the importance to cognitive psychologists of developing a path for the transfer of re-search findings to address real-world concerns. Each individual researcher will need to take advantage of opportunities to interact with usability engineers to (a) discover their problems and concerns related to usability methods and (b) relate basic cogni-tive and perceptual findings and principles to their specific design problems (a good example of this kind of activity can be found in Wickens, Vincow, Schopper, & Lin-coln, 1997). One implication is that usability scientists need to communicate in fo-rums used by usability engineers, not just those used by other usability scientists or basic researchers. These communication outlets include the Usability Professionals Association's (UPA's) annual meeting (the Web site URL for the UPA is http://www.upassoc.org) and the various electronic mailing list forums on the World Wide Web. Unfortunately, the rewards for academic researchers communicat-ing in such forums may not come from the organization that employs the scientists (see Gillan & Bias, 1992, for a more detailed discussion of the role of rewards in bridg-ing the gap between basic research and usability design). This introduces the second type of change that will be necessary for usability science to succeed: Organizations such as university departments and colleges, funding agencies such as the National Science Foundation, and corporations that employ usability engineers will need to be convinced of the value of usability science. One approach is to develop small-scale proof-of-concept projects that can be used to convince existing organizations of the value of translating basic research findings into design principles (see Bias, Gillan, & Tullis, 1993; Gillan & Bias, 1992, for examples). Gillan and Schvaneveldt (1999) sug-gested that building bridges between different disciplines is hard work; changing cultures, as this final point suggests, is even harder work. The value of the goals—better usability engineering, a cognitive psychology that can affect people's lives in concrete, positive ways, and more usable technology—makes it worth the effort.

REFERENCES

Anderson, J. R. (1983). *The architecture of cognition.* Cambridge, MA: Harvard University Press.
Anderson, J. R. (1993). *Rules of the mind.* Hillsdale, NJ: Lawrence Erlbaum Associates, Inc.

Angell, J. R. (1907). The province of functional psychology. *Psychological Review, 14,* 61–91.

Bailey, G. (1993). Iterative methodology and designer training in human–computer interface design. In S. Ashlund, K. Mullet, A. Henderson, E. Hollnagel, & T. White (Eds.), *Proceedings of INTERCHI '93 Conference on Human Factors in Computing Systems* (pp. 24–29). New York: ACM.

Bias, R. G., & Gillan, D. J. (1998). Whither the science of human–computer interaction? A debate involving researchers and practitioners. In *Proceedings of the Human Factors and Ergonomics Society 42nd Annual Meeting* (p. 526). Santa Monica, CA: Human Factors and Ergonomics Society.

Bias, R. G., Gillan, D. J., & Tullis, T. S. (1993). Three usability enhancements to the human factors-design interface. In G. Salvendy & M. J. Smith (Eds.), *Proceedings of the Fifth International Conference on Human–Computer Interaction: HCI International '93* (pp. 169–174). Amsterdam: Elsevier.

Broadbent, D. E. (1980). The minimization of models. In A. Chapman & D. Jones (Eds.), *Models of man* (pp. 113–128). Leicester, England: British Psychological Society.

Brock, T. D. (1999). *Robert Koch: A life in medicine and bacteriology.* Washington, DC: ASM Press.

Card, S. K., English, W. K., & Burr, B. J. (1978). Evaluation of mouse, rate-controlled isometric joystick, and text keys for text selection on a CRT. *Ergonomics, 21,* 601–613.

Card, S. K., Moran, T. P., & Newell, A. (1983). *The psychology of human–computer interaction.* Hillsdale, NJ: Lawrence Erlbaum Associates, Inc.

Carr, H. A. (1925). *Psychology: A study of mental activity.* New York: Longmans Green.

Carroll, J. M. (Ed.). (1991). *Designing interaction: Psychology at the human–computer interaction.* New York: Cambridge University Press.

Carroll, J. M. (1997). Human–computer interaction: Psychology as a science of design. *International Journal of Human–Computer Studies, 46,* 501–522.

Carroll, J. M., & Campbell, R. L. (1989). Artifacts as psychological theories: The case for human–computer interaction. *Behavior and Information Technology, 8,* 247–256.

Carroll, J. M., & Kellogg, W. A. (1989). Artifact as theory-nexus: Hermeneutics meets theory-based design. In K. Bice & C. Lewis (Eds.), *Proceedings of SIGCHI '89 Conference on Human Factors in Computing Systems* (pp. 7–14). New York: ACM.

Carswell, C. M. (1992). Reading graphs: Interactions of processing requirements and stimulus structure. In B. Burns (Ed.), *Percepts, concepts, and categories* (pp. 605–645). Amsterdam: Elsevier.

Carswell, C. M., & Wickens, C. D. (1988). *Comparative graphics: History and applications of perceptual integrality theory and the proximity compatibility hypothesis.* (University of Illinois Tech. Rep. No. ARL–88–2/AHEL–88–1/AHEL, Technical Memorandum 8–88). Savoy, IL: Aviation Research Laboratory.

Cook, T. D., & Campbell, D. T. (1979). *Quasi-experimentation: Design and analysis studies issues for field studies.* Chicago: Rand McNally.

Cooke, N. M., & Gillan, D. J. (1999). Representing human behavior in human–computer interaction. In *Encyclopedia of computer science and technology* (Vol. 40, Suppl. 25, pp. 283–308). New York: Marcel Dekker.

Cooke, N. M., Neville, K. J., & Rowe, A. L. (1996). Procedural network representations of sequential data. *Human–Computer Interaction, 11,* 29–68.

Cooke, N. M., Stout, R., Rivera, K., & Salas, E. (1998). Exploring measures of team knowledge. In *Proceedings of the Human Factors and Ergonomics Society 42nd Annual Meeting* (pp. 215–219). Santa Monica, CA: Human Factors and Ergonomics Society.

Cooper, L. A. (1979). Individual differences in visual comparison processes. *Perception and Psychophysics, 19,* 433–444.

Desuvire, H. W., Kondziela, J. M., & Atwood, M. E. (1992). What is gained and lost when using evaluation methods other than empirical testing. In *Proceedings of the HCI '92 Conference on People and Computers VII* (pp. 89–102). New York: Cambridge University Press.

Dewey, J. (1896). The reflex arc concept in psychology. *Psychological Review, 3*, 357–370.

Fodor, J. (1983). *The modularity of mind*. Cambridge, MA: MIT Press/Bradford Books.

Gardner, H. (1972). *The quest for mind: Piaget, Levi-Strauss, and the structuralist movement*. New
York: Vintage.

Gillan, D. J. (1995). Visual arithmetic, computational graphics, and the spatial metaphor. *Human Factors, 37*, 766–780.

Gillan, D. J. (1998, April). Whither the science of human–computer interaction? *CSTG Bulletin, 25*(2), 17–18.

Gillan, D. J. (2000a). A componential model of human interaction with graphs. V. Using pie
graphs to make comparisons. In *Proceedings of Human Factors and Ergonomics Society 44th
Annual Meeting* (pp. 3.439–3.442). Santa Monica, CA: Human Factors and Ergonomics Society.

Gillan, D. J. (2000b). Usability methods at the millennium: How we got here and where we
might be going. In *Proceedings of the Human Factors and Ergonomics Society* (pp. 1.315–
1.318). Santa Monica, CA: Human Factors and Ergonomics Society.

Gillan, D. J., & Bias, R. G. (1992). The interface between human factors and design. *Proceedings
of the Human Factors Society 36th Annual Meeting* (pp. 443–447). Santa Monica, CA: Human
Factors and Ergonomics Society.

Gillan, D. J., & Callahan, A. B. (2000). A componential model of human interaction with
graphs. VI. Cognitive engineering of pie graphs. *Human Factors, 42*, 566–591.

Gillan, D. J., & Cooke, N. M. (1998). Making usability data more usable. In *Proceedings of Human Factors and Ergonomics Society 42nd Annual Meeting* (pp. 300–304). Santa Monica, CA:
Human Factors and Ergonomics Society.

Gillan, D. J., & Cooke, N. M. (2000). Using pathfinder networks to analyze procedural knowledge in interactions with advanced technology. In E. Salas (Ed.), *Human–technology interaction in complex systems* (Vol. 10). Greenwich, CT: JAI.

Gillan, D. J., & Harrison, C. (1999). A componential model of human interaction with graphs.
IV. Holistic and analytical perception of star graphs. In *Proceedings of Human Factors and Ergonomics Society 43rd Annual Meeting* (pp. 1304–1307). Santa Monica, CA: Human Factors
and Ergonomics Society.

Gillan, D. J., Holden, K. L., Adam, S., Rudisill, M., & Magee, L. (1990). How does Fitts' law fit
pointing and dragging? In J. C. Chew & J. Whiteside (Eds.), *Proceedings of CHI '90 Conference on Human Factors in Computing Systems* (pp. 227–234). New York: ACM.

Gillan, D. J., Holden, K., Adam, S., Rudisill, M., & Magee, L. (1992). How should Fitts' law be
applied to human–computer interaction? *Interacting With Computers, 4*, 289–313.

Gillan, D. J., & Lewis, R. (1994). A componential model of human interaction with graphs. I.
Linear regression modeling. *Human Factors, 36*, 419–440.

Gillan, D. J., & Neary, M. (1992). A componential model of interaction with graphs. II. The effect of distance between graphical elements. In *Proceedings of the Human Factors Society 36th
Annual Meeting* (pp. 365–368). Santa Monica, CA: Human Factors and Ergonomics Society.

Gillan, D. J., & Schvaneveldt, R. W. (1999). Applying cognitive psychology: Bridging the gulf
between basic research and cognitive artifacts. In F. T. Durso, R. Nickerson, R. Schvaneveldt, S. Dumais, M. Chi, & S. Lindsay (Eds.), *The handbook of applied cognition* (pp. 3–31).
Chichester, England: Wiley.

Gillan, D. J., Wickens, C. D., Hollands, J. G., & Carswell, C. M. (1998). Guidelines for presenting quantitative data in Human Factors and Ergonomics Society publications. *Human Factors, 40*, 28–41.

Gray, W. D., & Salzman, M. C. (1998). Damaged merchandise? A review of experiments that compare usability evaluation methods. *Human–Computer Interaction, 13,* 203–261.

Hempel, C. G. (1966). *Philosophy of natural science.* Englewood Cliffs, NJ: Prentice Hall.

Human Factors and Ergonomics Society. (2000). *HFES graduate program directory.* Santa Monica, CA: Author.

Inkpen, K., Booth, K. S., & Klawe, M. (1996). Interaction styles for educational computer environments: A comparison of drag-and-drop versus point-and-click. (Tech. Rep. No. 96–17). Vancouver, Canada: University of British Columbia, Department of Computer Science.

James, W. (1907). *Pragmatism: A new name for some old ways of thinking.* New York: Longmans Green.

James, W. (1975). Philosophical and practical results. In F. Burkhardt (Ed.), *Pragmatism* (pp. 257–270). Cambridge, MA: Harvard University Press.

Jeffries, R., Miller, J. R., Wharton, C., & Uyeda, K. M. (1991). User interface evaluation in the real world: A comparison of four techniques. In S. P. Robertson, G. M. Olson, & J. S. Olson (Eds.), *Proceedings of CHI '90 Conference on Human Factors in Computing Systems* (pp. 119–124). New York: ACM.

Karat, C.-M., Campbell, R., & Fiegel, T. (1992). Comparison of empirical testing and walk-through methods in user interface evaluation. In P. Bauersfield, J. Bennett, & G. Lynch (Eds.), *Proceedings of CHI '92 Conference on Human Factors in Computing Systems* (pp. 397–404). New York: ACM.

Karat, C.-M., Lund, A., Coutaz, J., & Karat, J. (Eds.). (1998). *Proceedings of CHI '98 Conference on Human Factors in Computing Systems.* New York: ACM.

Katz, I. R., Mack, R., Marks, L., Rosson, M. B., & Nielsen, J. (Eds.). (1995). *Proceedings of CHI '95 Conference on Human Factors in Computing Systems.* New York: ACM.

Landauer, T. K. (1987). Relations between cognitive psychology and computer system design. In J. M. Carroll (Ed.), *Interfacing thought* (pp. 1–25). Cambridge, MA: MIT Press.

Lohse, G. L. (1993). A cognitive model for understanding graphical perception. *Human–Computer Interaction, 8,* 313–334.

Long, J. (1996). Specifying relations between research and the design of human–computer interactions. *International Journal of Human–Computer Studies, 44,* 875–920.

MacKenzie, I. S. (1992). Fitts' law as a research and design tool in human–computer interaction. *Human–Computer Interaction, 7,* 91–139.

MacKenzie, I. S., Sellen, A., & Buxton, W. (1991). A comparison of input devices in elemental pointing and dragging tasks. In S. P. Robertson, G. M. Olson, & J. S. Olson (Eds.), *Proceedings of CHI '91 Conference On Human Factors in Computing Systems* (pp. 161–166). New York: ACM.

McClelland, J. L., & Rumelhart, D. E. (Eds.). (1986). *Parallel distributed processing: Explorations in the microstructure of cognition* (Vol. 2). Cambridge, MA: MIT Press/Bradford Books.

Meister, D. (1989). *The conceptual aspects of human factors.* Baltimore: Johns Hopkins University Press.

Neel, A. (1969). *Theories of psychology: A handbook.* London: University of London Press.

Neisser, U. (1967). *Cognitive psychology.* Englewood Cliffs, NJ: Prentice Hall.

Newell, A., & Card, S. K. (1985). The prospects for psychological science in human–computer interaction. *Human–Computer Interaction, 1,* 209–242.

Nielsen, J. (1992). Finding usability problems through heuristic evaluation. In P. Bauersfield, J. Bennett, & G. Lynch (Eds.), *Proceedings of CHI '92 Conference on Human Factors in Computing Systems* (pp. 373–380). New York: ACM.

Nielsen, J. (1999). The top ten *new* mistakes of Web design. *Alertbox, May 30, 1999.* Retrieved June 17, 1999, from http://www.useit.com/alertbox/990530.html

Nielsen, J., & Phillips, V. L. (1993). Estimating the relative usability of two interfaces: Heuristic, formal, and empirical methods. In S. Ashlund, K. Mullet, A. Henderson, E. Hollnagel, & T. White (Eds.), *Proceedings of INTERCHI '93 Conference on Human Factors in Computing Systems* (pp. 214–221). New York: ACM.

Norman, D. A. (1987). Cognitive engineering—cognitive science. In J. M. Carroll (Ed.), *Interfacing thought* (pp. 325–336). Cambridge, MA: MIT Press.

Norman, D. A. (1988). *The psychology of everyday things.* New York: Basic Books.

Olson, J. S., & Olson, G. M. (1999). Computer-supported cooperative work. In F. T. Durso, R. Nickerson, R. Schvaneveldt, S. Dumais, M. Chi, & S. Lindsay (Eds.), *The handbook of applied cognition* (pp. 409–442). Chichester, England: Wiley.

Peirce, C. S. (1878). How to make our ideas clear. *Popular Science Monthly, 12,* 286–302. (Reprinted in *Pragmatism: A reader,* by L. Menand, Ed., 1997, New York: Vintage)

Rhodes, R. (1997). *Deadly feasts.* New York: Simon & Schuster.

Simkin, D., & Hastie, R. (1987). An information processing analysis of graph perception. *Journal of the American Statistical Association, 82,* 454–465.

Squire, L. R., Knowlton, B., & Musen, G. (1993). The structure and organization of memory. *Annual Review of Psychology, 44,* 453–495.

Sternberg, R. J. (1999). A dialectic basis for understanding the study of cognition. In R. J. Sternberg (Ed.), *The nature of cognition* (pp. 51–78). Cambridge, MA: MIT Press.

Titchener, E. B. (1898). The postulates of a structural psychology. *Philosophical Review, 7,* 449–465.

Tulving, E. (1985). How many memory systems are there? *American Psychologist, 40,* 385–398.

Tversky, B., Kugelmass, S., & Winter, A. (1991). Cross-cultural and developmental trends in graphic productions. *Cognitive Psychology, 23,* 515–557.

Walker, N., & Smelcer, J. B. (1990). A comparison of selection times from walking and pull-down menus. In J. C. Chew & J. Whiteside (Eds.), *Proceedings of CHI '90 Conference on Human Factors in Computing Systems* (pp. 221–225). New York: ACM.

Weber, R. J. (1992). *Forks, phonographs, and hot air balloons.* New York: Oxford University Press.

Wickens, C. D., & Carswell, C. M. (1995). The proximity compatibility principle: Its psychological foundations and relevance to display design. *Human Factors, 37,* 473–494.

Wickens, C. D., Vincow, M. A., Schopper, A. A., & Lincoln, J. E. (1997). Computational models of human performance in the design and layout of displays and controls. In *CSERIAC State of the Art Report (SOAR).* Dayton, OH: Wright–Patterson AFB, Crew Systems Information Analysis Center.

Wilcox, S. B. (1992). Functionalism then and now. In D. A. Owens & M. Wagner (Eds.), *Progress in modern psychology: The legacy of American functionalism* (pp. 31–51). Westport, CT: Praeger.

Woods, D. (1998). Designs are hypotheses about how artifacts shape cognition and collaboration. *Ergonomics, 41,* 168–173.

INTERNATIONAL JOURNAL OF HUMAN–COMPUTER INTERACTION, *13*(4), 373–410
Copyright © 2001, Lawrence Erlbaum Associates, Inc.

Criteria For Evaluating Usability Evaluation Methods

H. Rex Hartson
Department of Computer Science
Virginia Tech

Terence S. Andre
Air Force Research Laboratory

Robert C. Williges
Department of Industrial and Systems Engineering
Virginia Tech

The current variety of alternative approaches to usability evaluation methods (UEMs) designed to assess and improve usability in software systems is offset by a general lack of understanding of the capabilities and limitations of each. Practitioners need to know which methods are more effective and in what ways and for what purposes. However, UEMs cannot be evaluated and compared reliably because of the lack of standard criteria for comparison. In this article, we present a practical discussion of factors, comparison criteria, and UEM performance measures useful in studies comparing UEMs. In demonstrating the importance of developing appropriate UEM evaluation criteria, we offer operational definitions and possible measures of UEM performance. We highlight specific challenges that researchers and practitioners face in comparing UEMs and provide a point of departure for further discussion and refinement of the principles and techniques used to approach UEM evaluation and comparison.

1. INTRODUCTION

The concept of evaluation dates back to the beginning of system analysis and human factors and beyond. Usability evaluation reaches back to virtually the beginning of human–computer interaction (HCI). Usability evaluation methods (UEMs) go back more than 2 decades, published accounts of UEMs go back more than a decade (Card, Moran, & Newell, 1983; Nielsen & Molich, 1990), and for some time researchers have conducted studies that compare UEMs (Jeffries, Miller, Wharton, &

Thanks to the anonymous reviewer who suggested including an adaptation of the *F* measure for information retrieval in our section on effectiveness.

Requests for reprints should be sent to H. Rex Hartson, Department of Computer Science–0106, Virginia Technological Institute, Blacksburg, VA 24061. E-mail: hartson@vt.edu

Uyeda, 1991; Nielsen & Molich, 1990). However, in a broad historical view, the area is still relatively new and incomplete as both a research topic and as an applied body of knowledge.

In the 1980s, laboratory usability testing quickly became the primary UEM for examining a new or modified interface. Laboratory usability testing was seen by developers as a way to minimize the cost of service calls, increase sales through the design of a more competitive product, minimize risk, and create a historical record of usability benchmarks for future releases (Rubin, 1994). Laboratory testing involved user performance testing to evaluate speed, accuracy, and errors in addition to user subjective evaluations. Methods for collecting data beyond user performance data included verbal protocols (Ericsson & Simon, 1984), critical incident reporting (del Galdo, Williges, Williges, & Wixon, 1987), and user satisfaction ratings (Chin, Diehl, & Norman, 1988). In the 1990s, many developers explored other methods in an attempt to bring down the cost and time requirements of traditional usability testing. In addition, because usability testing often had been occurring late in the design process, developers were motivated to look at methods that could be used earlier when only an immature design was available (Marchetti, 1994). As a result, expert-based inspection methods grew in popularity because many of them were intended to be used with a relatively early design concept (Bradford, 1994). Some of the more popular expert-based UEMs include guideline reviews based on interaction design guidelines such as those by Smith and Mosier (1986), heuristic evaluation (Nielsen & Molich, 1990), cognitive walk-throughs (C. Lewis, Polson, Wharton, & Rieman, 1990; Wharton, Bradford, Jeffries, & Franzke, 1992), usability walk-throughs (Bias, 1991), formal usability inspections (Kahn & Prail, 1994), and heuristic walk-throughs (Sears, 1997).

Practitioners are far from settled on a uniform UEM, and researchers are far from agreement on a standard means for evaluating and comparing UEMs. Confounding the situation is a miscomprehension of the limitations of UEMs and to what conditions those limitations apply. What makes the state of UEM affairs most disturbing to many is the lack of understanding of UEM evaluation and comparison studies, as pointed out by Gray and Salzman (1998) and debated in subsequent discussions (Olson & Moran, 1998). Gray and Salzman called into question existing UEM studies as potentially misleading and urged researchers to apply the power of the experimental method more carefully and rigorously in these studies.

1.1. Iterative Design and Evaluation

Interactive systems, at least the user interfaces, are usually designed through an iterative process involving design, evaluation, and redesign. Kies, Williges, and Rosson (1998) summarized three major iterative stages of initial, prototype, and final design that are central to the iterative design process. During initial design, goals and guidelines are iterated to finalize the design specifications leading to a prototype design. Formative evaluation focuses on usability problems that need to be solved during the prototype design stage before a final design can be accepted for release. Summative evaluation is then conducted to evaluate the efficacy of the final design or to compare competing design alternatives in terms of usability. As shown in Figure 1, UEMs are used primarily for formative evaluations during the prototype design stage. These formative evaluations are focused on efficient and effective techniques to determine usability problems that need to be eliminated through redesign. A combination of expert-based and user-based inspection methods has evolved to facilitate the formative evaluation process.

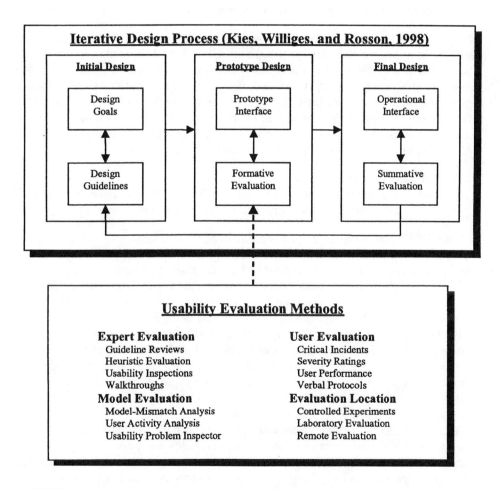

FIGURE 1 UEMs used in formative usability evaluation. From "Coordinating computer-supported cooperative work: A review of research issues and strategies" by J. K. Kies, R. C. Williges, and M. B. Rosson, 1998, *Journal of the American Society for Information Science, 49,* pp. 776–779. Copyright 1994 by John Wiley & Sons. Reprinted with permission.

1.2. Need for a Foundation for Evaluating Usability Evaluation Methods

Among interactive system developers and users there is now much agreement that usability is an essential quality of software systems. Among the HCI and usability communities, there is also much agreement that

- Usability is seated in the interaction design.
- An iterative, evaluation-centered process is essential for developing high usability in interaction designs.
- A class of usability techniques called *UEMs* have emerged to support that development process by evaluating the usability of interaction designs and identifying usability problems to be corrected.

Beyond this level of agreement, however, there are many ways to evaluate the usability of an interaction design (i.e., many UEMs), and there is much room for disagreement and discussion about the relative merits of the various UEMs. As more new methods are being introduced, the variety of alternative approaches and a general lack of understanding of the capabilities and limitations of each has intensified the need for practitioners and others to be able to determine which methods are more effective and in what ways and for what purposes. In reality, researchers find it difficult to reliably compare UEMs because of a lack of

- Standard criteria for comparison.
- Standard definitions, measures, and metrics on which to base the criteria.
- Stable, standard processes for UEM evaluation and comparison.

Lund (1998) noted the need for a standardized set of usability metrics, citing the difficulty in comparing various UEMs and measures of usability effectiveness. As Lund pointed out, there is no single standard for direct comparison, resulting in a multiplicity of different measures used in the studies, capturing different data defined in different ways. Consequently, very few studies clearly identify the target criteria against which to measure success of a UEM being examined. As a result, the body of literature reporting UEM comparison studies does not support accurate or meaningful assessment or comparisons among UEMs. Many such studies that have been reported were not complete or otherwise fell short of the kind of scientific contribution needed. Although these shortcomings often stemmed from practical constraints, they have led to substantial critical discussion in the HCI literature (Gray & Salzman, 1998; Olson & Moran, 1998).

Accordingly, in this article we present a practical discussion of factors, comparison criteria, and UEM performance measures that are interesting and useful in studies comparing UEMs. We highlight major considerations and concepts, offering some operational definitions and exposing the hazards of some approaches proposed or reported in the literature. In demonstrating the importance of developing appropriate UEM evaluation criteria, we present some different possible measures of effectiveness, select and review studies that use two of the more popular measures, and consider the trade-offs among different criterion definitions. This work highlights some of the specific challenges that researchers and practitioners face when comparing UEMs and provides a point of departure for further discussion and refinement of the principles and techniques used to approach UEM evaluation and comparison.

1.3. Terminology

As popularized by Gray and Salzman (1998), we use the term *usability evaluation method* (UEM) to refer to any method or technique used to perform formative usability evaluation (i.e., usability evaluation or testing to improve usability) of an interaction design at any stage of its development. This broad definition includes laboratory-based formative usability testing with users, heuristic and other expert-based usability inspection methods, model-based analytic methods, all kinds of expert evaluation, and remote evaluation of interactive software after deployment in the field. As discussed earlier, we exclude summative studies of usability in a given product from the concept of a UEM. We use the term *UEM comparison study* to refer to any empirical summative evaluation that compares performance (by any measure) among UEMs.

The essential common characteristic of UEMs (at least for purposes of this article) is that every UEM, when applied to an interaction design, produces a list of potential usability problems as its output. Some UEMs have additional functionality, such as the ability to help write usability problem reports, to classify usability problems by type, to map problems to causative features in the design, or to offer redesign suggestions. We believe these are all important and deserve attention in addition to the basic performance-based studies.

A person using a UEM to evaluate usability of an interaction design is called an *evaluator,* to distinguish this specialized usability engineering role. More specifically, a person using a usability inspection method (one type of UEM) is often called an *inspector.*

1.4. Types of Evaluation and Types of UEMs

In the previous sections, we have used the terms *formative evaluation* and *summative evaluation.* To understand UEMs and their evaluation, one must understand evaluation in these terms and in the context of usability. We have adopted Scriven's (1967) distinction between two basic approaches to evaluation based on the evaluation objective. Formative evaluation is evaluation done during development to improve a design, and summative evaluation is evaluation done after development to assess a design (absolute or comparative). Phrasing Scriven's definitions in terms of usability, formative evaluation is used to find usability problems to fix so that an interaction design can be improved. Summative evaluation is used to assess or compare the level of usability achieved in an interaction design. Summative evaluation is generally regarded as requiring rigorous, formal experimental design, including a test for statistical significance and is often used to compare design factors in a way that can add to the accumulated knowledge within the field of HCI. Although the words *usability evaluation method,* taken at face value, technically could include formal methods for controlled empirical studies of usability, the convention is to limit the term *UEM* to refer to methods for formative usability evaluation. Further, because this article is about summative studies of UEMs, classifying summative usability studies as UEMs would necessitate a discussion about summative studies of summative studies, which is beyond the scope of this article. Thus, we limit the scope of this article to UEMs used to perform formative usability evaluation of interaction designs and do not include summative studies of usability of a system as a UEM.

Usually, formative UEMs are associated with qualitative usability data (e.g., usability problem identification) and summative usability evaluation with quantitative data (e.g., user performance numbers). Sometimes formative usability evaluation can also have a component with a summative flavor. To approximate the level of usability achieved, some UEMs lend an informal (not statistically rigorous) summative flavor to the formative process by supporting collection of quantitative usability data (e.g., time on task) in addition to the qualitative data. Not being statistically significant, these results do not contribute (directly) to the science of usability but are valuable usability engineering measures within a development project. Usability engineers, managers, and marketing people use quantitative usability data to identify convergence of a design to an acceptable level of usability and to decide (as an engineering or management decision, not a scientific decision) when to stop iterating the development process. Beyond this summative flavor, however, UEMs are about qualitative usability data, not quantitative data. This issue arises again, in Section 7.1, in terms of focusing on qualitative data of UEMs in comparison studies of UEMs.

A somewhat orthogonal perspective is used to distinguish evaluation methods in terms of how evaluation is done. Hix and Hartson (1993b) described two kinds of evaluation: analytic and empirical. Analytic evaluation is based on analysis of the characteristics of a design through examination of a design representation, prototype, or implementation. Empirical evaluation is based on observation of performance of the design in use. Perhaps Scriven (1967), as described by Carroll, Singley, and Rosson (1992), gets at the essence of the differences better by calling these types of evaluation, respectively, *intrinsic evaluation* and *payoff evaluation*.[1] Intrinsic evaluation is accomplished by way of an examination and analysis of the attributes of a

---

[1] In describing Scriven's (1967) distinction between intrinsic and payoff approaches to evaluation, other authors (e.g., Carroll et al., 1992; Gray & Salzman, 1998) quoted his example featuring an ax:

If you want to evaluate a tool, say an axe, you might study the design of the bit, the weight distribution, the steel alloy used, the grade of hickory in the handle, etc., or you might just study the kind and speed of the cuts it makes in the hands of a good axeman. (Scriven, 1967, p. 53)

Although this example served Scriven's (1967) purpose well, it also offers us a chance to make a point about the need to carefully identify usability goals before establishing evaluation criteria. Giving an HCI usability perspective to the ax example, we see that user performance observation in payoff evaluation does not necessarily require an expert axman (or axperson). Expert usage might be one component of the vision in ax design for usability, but it is not an essential part of the definition of payoff evaluation. Usability goals depend on expected user classes and the expected kind of usage. For example, an ax design that gives optimum performance in the hands of an expert might be too dangerous for a novice user. For the city dweller, also known as weekend wood whacker, safety might be a usability goal that transcends firewood production, calling for a safer design that might necessarily sacrifice efficiency. One hesitates to contemplate the metric for this case, possibly counting the number of 911 calls from a cell phone in the woods. Analogously in the user interface domain, usability goals for a novice user of a software accounting system, for example, might place data integrity (error avoidance) above sheer productivity.

design without actually putting the design to work, whereas payoff evaluation is evaluation situated in observed usage.

The de facto standard payoff method in the usability world is the well-known laboratory-based usability testing with users. Goals, operators, methods, selection (Card et al., 1983) analysis—in which user actions for task performance are assigned costs (in terms of time), set within a model of the human as information processor—offers a good representative example of intrinsic usability evaluation. Some usability inspection methods (Nielsen & Mack, 1994) are essentially intrinsic in that they analyze an interaction design with respect to a set of design guidelines or heuristics. This kind of inspection method requires a usability expert to analyze the design rather than to test it with real users. Other usability inspection methods are hybrids between intrinsic and payoff in that the analysis done during usability inspection is task driven; the expert's analysis is based on exploring task performance and encountering usability problems in much the same way users would, adding a payoff dimension to the intrinsic analysis. In this situation, a usability inspector asks questions about designs in the context of tasks to predict problems users would have.

Regardless of the method, the goal of all UEMs is essentially the same: to produce descriptions of usability problems observed or detected in the interaction design for analysis and redesign. Ostensibly, this shared goal and common output should make the various UEMs directly comparable, but as we already know from the discussion in the literature, things are not that simple.

1.5. Damaged Merchandise

The need to evaluate and compare UEMs is underscored by the fact that some developers have recently questioned the effectiveness of some types of UEMs in terms of their ability to predict problems that users actually encounter (John & Marks, 1997). Gray and Salzman (1998) recently documented, in their article about "damaged merchandise," specific validity concerns about five popular UEM comparison studies. The term *damaged merchandise* is a reference to the lack of attention given by researchers to rigorous experimental design for evaluating UEMs. Gray and Salzman made the case that, when the results of a study not rigorously designed and executed according to the prescripts of experimental design methodology for statistical significance are used to inform the choice of which UEM to use, the consequence is damaged merchandise. They made the point that even small problems with experimental studies call into question what is accepted as known about UEMs. A key concern noted by Gray and Salzman is the issue of using the right measure (or measures) to compare UEMs in terms of effectiveness.

To be fair, some of the incomplete results criticized by Gray and Salzman (1998) were understandable because researchers were using data that became available through means designed for other ends (e.g., usability testing within a real development project), and additional resources were not available to conduct a complete, scientifically valid experiment. It is fair to say that these partial results have value as indicators of relative UEM merit in a field in which complete scientific results are scarce. In many engineering contexts, usability engineering included, so-called damaged merchandise is not always a bad thing. For example, most UEMs represent a conscious trade-off of performance for savings in cost. As long as "buyers" know what they are getting, and it will suffice for their needs, they can often get a good "price" for damaged merchandise. This is, we think, a sound principle behind what has been called *discount engineering* (Nielsen, 1989), and it has always been part of the legitimate difference between science and engineering.

However, as pointed out by Gray and Salzman (1998), damaged merchandise is far less acceptable in the realm of usability science, especially when found in the form of poorly designed UEM comparison studies. There is certainly a need for more carefully designed comparison experiments—both to contribute to the science of usability and to provide practitioners with more reliable information about the relative performance of various UEMs as used for various purposes. Some authors in the discussion sequel to the Gray and Salzman article, as compiled by Olson & Moran (1998), have suggested that some science is better than none, and resource limitations that preclude complete scientific results should not prevent attempts at modest contributions. These discussants have argued that this is especially true in a relatively new field in which any kind of result is difficult to come by. In balance, Gray and Salzman would probably caution us that sometimes bad science is worse than none. However, as Lund (1998) pointed out in his commentary about Gray and Salzman, the danger may not loom so darkly to practitioners, making the case that practitioners will quickly discover if a recommendation is not useful. In any case, the argument made previously, which applies to the acceptance of any merchandise, was based on the buyers knowing what they are, and are not, getting for their money.

1.6. Road Map of Concepts

Figure 2 shows a guide to the concepts of this article and the relations among them. Researchers planning a UEM comparison study have in mind an ultimate criterion for establishing the "goodness" of a particular method. However, because ultimate criteria are usually considered impossible to measure directly, researchers select one of many possible actual criteria to approximate the ultimate criterion for UEM comparison. The experimenter then applies a method representing the actual criterion to identify a standard set of usability problems existing in the target system interaction design. The experimenter also applies the UEMs being compared (UEM–A and UEM–B in Figure 2) to the target design and calculates UEM performance metrics using the resulting usability problem lists in relation to the standard usability problem list. The UEMs are then compared on the basis of their performance metrics computed from the problem lists. We discuss these concepts in detail in the sections that follow.

2. EVALUATION CRITERION RELEVANCE, DEFICIENCY, AND CONTAMINATION

To evaluate the effectiveness of a UEM, and especially to compare the effectiveness of UEMs, usability researchers must establish a definition for effectiveness and an evaluation or comparison criterion or criteria. The criteria are stated in terms of one or more performance-related (UEM performance, not user performance) measures

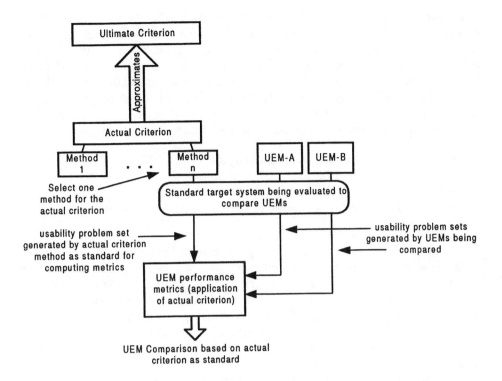

FIGURE 2   Roadmap of concepts.

(effectiveness indicators), which are computed from raw empirical usability data (e.g., usability problem lists) yielded by each UEM. Making the right choice for criteria and performance measures depends on understanding the alternatives available and the limitations of each. In this article, we bring these issues to light to foster this understanding.

The selection of criteria to evaluate a UEM is not essentially different from criteria selection for evaluation of other kinds of systems (Meister, Andre, & Aretz, 1997). In the evaluation of large-scale systems such as military weapon systems, for example, customers (e.g., the military commanders) establish ultimate criteria for a system in the real world. Ultimate criteria are usually simple and direct—for example, that a certain weapon system will win a battle under specified conditions. However, military commanders cannot measure such ultimate criteria directly outside of an actual combat environment. As a result, military commanders establish specific other attributes, called *actual criteria*, which are more easily measured and that there is reason to believe will be effective predictors of the ultimate criteria. To illustrate, commanders might establish the following characteristics as actual criteria for military aircraft performance: Aircraft must fly at X thousand feet, move at Y mach speed, and shoot with Z accuracy. As actual criteria, these measures are only indicators or predictors of the ultimate criterion and are more valuable as predictors if they can be validated, which can happen only when real combat clashes occur.

Measures to be used in actual criteria are designed to be operational parameters that can be computed by consistent means that are agreed on and reliably understood. If system reliability were a goal, for example, mean-time-between-failure would be a good measure because practitioners understand its meaning and computational mechanics. Researchers do not yet have any measures this well standardized in usability, so they generally define their own to meet the goals of the research (Gray & Salzman, 1998; Lund, 1998). To be useful and repeatable in an actual criterion, a measure must have at least these characteristics:

- A solid definition, understandable by all.
- A metric, to be computed from raw usability data.
- A standard way to measure or take data.
- One or more levels of performance that can be taken as a "score" to indicate goodness.

The degree to which actual criteria are successful predictors of the ultimate criterion is the essence of the concept called *criterion relevance,* illustrated by the intersection of the two circles in Figure 3. If, for example, stealth technology makes it unnecessary to fly at 80,000 feet, then the altitude criterion is no longer a useful predictor of the ultimate criterion causing that part of the actual criterion to fall outside the intersection with the ultimate criterion. Because this part of the actual criterion contaminates the approximation to the ultimate criterion, it is called *criterion contamination.*

If military commanders leave out an important measure that should be included in the estimate of an ultimate criterion, the actual criterion is deficient in representing the ultimate criterion, and the part of the ultimate criterion not represented falls outside the intersection in the part called *criterion deficiency.*

3. ULTIMATE CRITERION FOR UEM EFFECTIVENESS— FINDING REAL USABILITY PROBLEMS

Criterion relevance applies to UEMs as well as military planes. For discussion, we postulate the following ultimate criterion for UEM evaluation and compari-

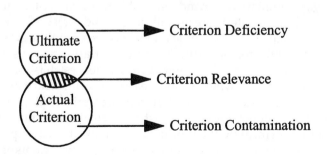

FIGURE 3   Relation between ultimate and actual criteria.

son, somewhat analogous to the simple ultimate criterion used in the case of the airplane: How well does the UEM help inspectors or evaluators discover real usability problems?

This *realness attribute,* which plays a pivotal role in several of the UEM measures is defined as follows: A usability problem (e.g., found by a UEM) is real if it is a predictor of a problem that users will encounter in real work-context usage and that will have an impact on usability (user performance, productivity, or satisfaction, or all three). This would exclude problems with trivially low impact and situations real users would or could not encounter. The emphasis on real users is important in this definition because many of the UEMs evaluated in studies are usability inspection methods in which the inspectors encounter problems that do not always predict usability problems for real users. In any case, this definition of realness belongs more to ultimate criteria than to any actual criterion because it does not yet offer an operational way to test for the stated conditions. This deceptively simple UEM criterion translates into a large number of issues when it comes to putting it into practice, when it comes to unpacking the meaning of the words *how well* and *real.*

To the extent that any practical means for determining realness in the actual criterion will result in some errors, there will be both criterion contamination and criterion deficiencies. However, once the actual criterion (including the method for determining realness) is established, those issues about "truth" regarding fidelity of the actual criteria to the ultimate criteria are encapsulated in the actual criteria selection process. This compartmentalization allows us to ignore questions of truth about the actual criteria during a study in which they are applied, where the focus is on the actual criterion as the standard. Here the question is how well does the UEM help inspectors discover real (as determined by the actual criterion) usability problems?

4. ACTUAL CRITERIA FOR UEM EFFECTIVENESS— OPERATIONALLY DETERMINING REALNESS

4.1. Determining Realness by Comparing With a Standard Usability Problem List

If an evaluator, or researcher, had a complete list of precisely the real usability problems that exist in a given target interaction design, that evaluator could ascertain the realness of each candidate usability problem found by a UEM. The evaluator would search the standard list for a match to the candidate problem, thereby determining whether it was in the list (and thus, whether it was real).

Usability problem lists as usability problem sets.   As we have established, each UEM, when applied, produces a list of usability problems. Because comparison of UEMs requires comparison and manipulation of their usability problem lists, it is often more useful to think of each UEM as producing a set of usability problems. As sets, the lists can be thought of as unordered, and they afford formal (set theoretic) expressions of important questions. Cockton and Lavery (1999) favored this same choice of terminology for much the same reasons. For example, one might need to ask whether a given UEM finds a certain known problem in a target design. Or one might need to know what usability problems the outputs of UEM_1 and UEM_2 have in common or what you get when you merge the outputs of UEM_1 and UEM_2. These are questions, respectively, about set membership, set intersections, and set unions. Simple set operations, such as union, intersection, and set difference can be used to manipulate the usability problems and combine them in various ways to calculate UEM performance measures.

Producing a standard usability problem set.   As a way to define realness, experimenters often seek to establish a standard touchstone set of usability problems deemed to be the real usability problems existing in the target interaction design of the study. This standard usability problem set is used as a basis for computing various performance measures as parts of actual criteria. We say that the touchstone set is part of an actual criterion because it can only approximate the theoretical ultimate real usability problem set, a set that cannot be computed. Some of the possible ways to produce a standard-of-comparison usability problem set for a given target interaction design include

- Seeding with known usability problems.
- Laboratory-based usability testing.
- Asymptotic laboratory-based testing.
- Union of usability problem sets over UEMs being compared.

The seeding approach introduces a known usability problem set to be used directly for comparison as part of the actual criterion. The two kinds of laboratory testing involve users and expert observers to produce standard usability problem sets found in the target system. The union of usability problem sets combines all the problem sets produced by the UEMs to produce the standard of comparison.

Seeding the target design with usability problems.   Sometimes experimenters will seed or "salt" a target system with known usability problems, an approach that can seem attractive because it gives control over the criterion. In fact, this is one of the few ways the experimenters can know about all the existing problems (assuming there are no real problems in the system before the seeding). Yet many UEM researchers believe salting the target system is not a good basis for the science of a UEM study because the outcome depends heavily on experimenter skill (in the salting), putting ecological validity in doubt. Experienced usability practitioners will know that contrived data can seldom match the variability, surprises, and realness of usability data from a usability laboratory.

Laboratory-based usability testing.   Traditional laboratory-based usability testing is the de facto standard, or the "gold standard" (Landauer, 1995; Newman, 1998) used most often in studies of UEM performance. Laboratory-based testing is a

UEM that produces high-quality, but expensive, usability problem sets. Often laboratory-based UEM performance is unquestioned in its effectiveness as a standard of comparison to evaluate other UEMs. Because it is such a well-established comparison standard, it might be thought of as an ultimate criterion, especially when compared to usability inspection methods. However, it does not meet our definition for an ultimate criterion because of the constraints and controls of the usability laboratory. Developers decide which tasks users should perform and what their work environment will be like (usually just the laboratory itself). Some researchers and practitioners would like more data on how well laboratory-based testing is predictive of real usability problems and under what conditions it best plays this role, but it is difficult to find an experimental standard good enough to make that comparison.

Despite these possible deviations from the ultimate, the experience of the usability community with laboratory-based testing as a mainstream UEM for formative evaluation within the interaction development process has led to a high level of confidence in this UEM. Other UEMs have arisen, not because of a search for higher quality but mostly out of a need for lower cost.

In any case, the typical laboratory-based usability test employs several users as participants along with one or more observers and produces the union of problems found by all users. Given that some usability problems, even from laboratory-based testing, can be of questionable realness, it is best to combine the laboratory test with expert review to eliminate some of the problems considered not real, thus improving the quality of the usability problem set to be used as the actual criterion.

Asymptotic laboratory-based testing. The typical usability laboratory test will miss some usability problems. In fact, most laboratory tests are deliberately designed with an objective of cost effectiveness at an acknowledged penalty of missing some usability problems. Using the formula $1 - (1-p)^n$, researchers have shown that a sample size of 5 participant evaluators (n) is sufficient to find approximately 80% of the usability problems in a system if the average individual detection rate (p) is at least 0.30 (Nielsen, 1994; Virzi, 1990, 1992; Wright & Monk, 1991). Virzi (1992) found average individual detection rates ranging from 0.32 to 0.42. However, J. R. Lewis (1994) found that average detection rates can be as low as 0.16 in office applications. Figure 4 shows the problem discovery likelihood when individual detection rates range between 0.15 and 0.45, using the formula $1 - (1-p)^n$. The rate at which problem detection approaches the asymptote varies significantly depending on the individual detection rate. Only 8 evaluators are needed to find 95% of the problems when the detection rate is 0.45, but as many as 19 evaluators are needed to find the same amount when the detection rate is 0.15.

For an individual detection rate of about 0.3 or higher, the first three to five users are enough to find 80% of the usability problems, as found independently by Nielsen (1990, 1992) and Virzi (1992). The number of new problems found by each added user levels off at about three to five users, with the number of new usability problems found dropping with each new user added after that. Therefore efficiency, or cost effectiveness, also levels off at three to five users. Fortunately for usability practice, both Nielsen (1992) and Virzi (1992) found that this kind of lab-

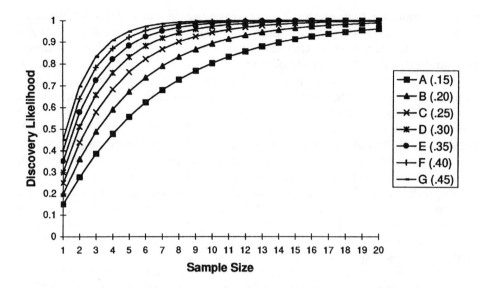

FIGURE 4   The asymptotic behavior of discovery likelihood as a function of the number of users. Adapted from "Sample sizes for usability studies: Additional considerations" by J. R. Lewis, 1994, *Human Factors, 36,* pp. 368–378. Copyright 1994 by the Human Factors and Ergonomics Society, all rights reserved. Reprinted with permission.

oratory-based testing also has a tendency to find the high-severity usability problems first. However, in J. R. Lewis's (1994) study of office applications, problem discovery rates were the same regardless of problem impact rating, raising the question of whether high-severity problems are always found first.

In any case, the total number of usability problems found does level off asymptotically as the number of users increases. This means that the asymptotic level can be thought of as a good approximation to the level of the ultimate criterion (after any nonreal problems are removed). Thus, extending the usual laboratory-based usability test to include several more users is a good, but expensive, choice for producing a standard usability problem set from the target design as part of an actual criterion.

Union of usability problem sets.   Another technique often used to produce a standard usability problem set as a criterion for being real is the union set of all the individual usability problem sets, as found by each of the methods being compared (Sears, 1997). This approach has the advantage in that it requires no effort beyond applying the UEMs being studied, but it has the drawback that it eliminates the possibility to consider validity as a UEM measure because the basis for metrics is not independent of the data. This drawback is explained further in Section 5.2 (Validity).

Comparing usability problem descriptions.   Gray and Salzman (1998) correctly criticized just counting usability problems for UEM measures without determining if some usability problems found overlap or duplicated others. A determination of overlap cannot be made, though, without an ability to compare usability problem descriptions. Determining realness by comparing with a standard usability problem set also requires comparison. Comparison requires complete, unambiguous usability problem descriptions that facilitate distinguishing different types of usability problems.

This comparison is straightforward in abstract sets in which each element is unambiguously identified by name or value. If $x \in A$ and $x \in B$, then the appearance of x in A is identical to its appearance in B. However, usability problem sets from UEMs are more difficult to compare because they involve enumerated sets in which elements are represented by narrative problem descriptions and elements, not by having a unique canonical identity.

Because usability problem descriptions are usually written in an ad hoc manner, expressed in whatever terms seem salient to the evaluator at the time the problem is observed, it is not unusual for two observers to write substantially different descriptions of the same problem. However, to perform set operations on usability problem sets, one needs the ability to determine when two different usability problem descriptions are referring to the same underlying usability problem. This kind of comparison of textual problem descriptions is usually done by expert judgment but is subject to much variability. There is a need for a standard way to describe usability problems, for a framework within which usability problem descriptions can be more easily and more directly compared. We are working on just such a framework, called the *User Action Framework* (Hartson, Andre, Williges, & van Rens, 1999).

4.2. Determining Realness by Expert Review and Judgment

Realness of usability problems can also be determined by review and judgment of experts in which each candidate usability problem is examined by one or more usability experts and determined by some guideline to be real or not. This technique can also have the effect of accumulating a standard list, if the judgment results can be saved and reused. This technique can also be combined with the techniques described in the following sections to filter their standard usability problem lists, ensuring that the results are, by this judgment, real.

Often designers of UEM studies find that the guidelines for realness to be used in expert judgment are too vague or general to be applied reliably, and the judgments can vary with the expert and other experimental conditions. This introduces the possibility of a bias causing the usability problem lists of each UEM to be judged differently. As an alternative, as we have described, experimenters seek a standard usability problem list as a single standard against which to compare each UEM's output. However, this approach also involves judgment when it comes to comparing each usability problem against the standard list.

4.3. Determining Realness by End-User Review and Judgment

Because usability is ultimately determined by the end user, not an expert evaluator, realness of problems needs to be established by the user. Specific UEM proce-

dures have been adopted to enhance realness criteria based on problems specified by real users.

Critical incidents.   Often, verbal protocols provided by the user do not provide succinct problem descriptions. Del Galdo, Williges, Williges, and Wixon
(1986) modified the critical incident technique described by Flanagan (1954) to allow identification of events or phenomena occurring during task performance
that are indicators of usability problems, which are captured during formative
evaluation. Critical incidents in usability evaluations have been used in conjunction with expert-based and user-based UEMs for collecting usability problems either in laboratory or in remote usability evaluations (Hartson, Castillo, Kelso,
Kamler, & Neale, 1996; Thompson & Williges, 2000). In addition, Neale, Dunlap,
Isenhour, and Carroll (2000) developed a collaborative critical incident procedure
that requires dialogue between the user and the expert evaluator to enrich the usability problem specification. Reliably detecting critical incidents and translating
them into clear, consistent, and comparable usability problem descriptions requires practitioner skill.

Severity ratings.   The concept of realness of a candidate usability problem
was introduced as a way to distinguish trivial usability problems from important
ones in UEM studies. Although this simple test of problem impact is necessary, it
is not sufficient. A usability problem judged to be real can still have either only
minor impact on user satisfaction or it might have show-stopping impact on user
task performance. To further discriminate among degrees of impact, practitioners
have extended the binary concept of realness into a range of possibilities called
severity levels. Severity thus becomes another measure of the quality of each usability problem found by a UEM, offering a guide for practitioners in deciding
which usability problems are most important to fix. The working assumption is
that high-severity usability problems are more important to find and fix than
low-severity ones. Thus, a UEM that detects a higher percentage of the high-severity problems will have more utility than a UEM that detects larger numbers of
usability problems—but ones that are mostly low-severity (even though all problems found might be real by the definition used). There are numerous schemes
for subjectively determining severity ratings for usability problems. Nielsen
(1994) is a representative example. Rubin (1994) used a criticality rating combining severity and probability of occurrence. Hix and Hartson (1993a) used cost-im-
portance analysis to prioritize problems for fixing.

5. UEM PERFORMANCE MEASURES—APPLYING ACTUAL CRITERIA

Bastien and Scapin (1995) identified three measures for examining an evaluation
method: thoroughness, validity, and reliability. Sears (1997) also pointed out these

same measures, giving them somewhat different operational definitions. These basic three measures are

- *Thoroughness:* Evaluators want results to be complete; they want UEMs to find as many of the existing usability problems as possible.
- *Validity:* Evaluators want results to be "correct"; they want UEMs to find only problems that are real.
- *Reliability:* Evaluators want results to be consistent; they want results to be independent of the individual performing the usability evaluation.

As a practical matter, we add a metric we call *effectiveness,* which is a combination of thoroughness and validity. Additionally, on behalf of practitioners who must get real usefulness within tightly constrained budgets and schedules, we also hasten to add *cost effectiveness* and *downstream utility* usefulness in the usability engineering process after gathering usability problem data (e.g., quality of usability problem reports in helping practitioners find solutions).

As Gray and Salzman (1998) pointed out, there is a need for multimeasure criteria, not just one-dimensional evaluations. When a researcher focuses on only one measure (e.g., thoroughness), it is unlikely that this one characteristic will reflect overall effectiveness of the UEM. In addition to thoroughness and validity, researchers may also be interested in reliability, cost effectiveness, downstream utility, and usability of UEMs. Any of these issues could form the criteria by which researchers judge effectiveness. Although it is nearly impossible to maximize all of the parameters simultaneously, practitioners must be aware that focusing on only one issue at the expense of others can lead to an actual criterion having significant criterion deficiency.

First, the ultimate criteria must be matched to the goals of evaluation. The main goal addressed by UEM evaluation is to determine which UEM is "best." Beyond that, we ask "Best for what?" Ultimate criteria should be selected with this more specific question in mind. In effectiveness studies of UEMs, the objective should then be to find those measures comprising actual criteria to best relate them to the ultimate criteria. Thus, the measures are a way of quantifying the question of how well a UEM meets the actual criteria.

5.1. Thoroughness

Thoroughness is perhaps the most attractive measure for evaluating UEMs. Thoroughness has a rather nice analogy to the concept of *recall* in the field of information storage and retrieval, a term that refers to a measure of retrieval performance of an information system from a target document collection (Salton & McGill, 1983). As an analogy, the document collection searched by an information system corresponds to the target interaction design being evaluated and the information retrieval system corresponds to the UEM. Documents found by an information system query correspond to usability problems found by a UEM. Recall is based on a concept called *relevance* (reflecting a determination of relevance of a document to a query), analogous to the concept of realness in UEMs. Relevance is the criterion for measuring precision and recall. Recall is a measure indicating the proportion of relevant documents found in a collection by an information system to the total relevant documents existing in the target document collection.

$$Recall = \frac{\text{number of relevant documents found}}{\text{number of relevant documents that exist}} \tag{1}$$

Analogously, Sears (1997) defined *thoroughness* as a measure indicating the proportion of real problems found using a UEM to the real problems existing in the target interaction design:

$$Thoroughness = \frac{\text{number of real problems found}}{\text{number of real problems that exist}} \tag{2}$$

For example, if a given UEM found only 10 of the 20 real usability problems that were determined to be in a target system (by some criterion yet to be discussed), that UEM would be said to have yielded a thoroughness of $10/20 = 0.5$. UEMs with low thoroughness leave important usability problems unattended after investment in the usability evaluation process.

Whatever method is used to determine realness, that method can also be considered a UEM, in this case a definitional UEM_A (*A* referring to "actual criteria") that, however arbitrarily, determines realness. The output of this so far undefined UEM is considered the "perfect" yardstick against which other UEMs are compared. When applied to the target interaction design, UEM_A produces a definitional usability problem set, *A*, defining (again, however arbitrarily) the real problems that exist in the design. If *P* is the set of usability problems detected by some UEM_P being evaluated, then the numerator for thoroughness of UEM_P is computed by an intersection with the standard as in this equation:

$$Thoroughness = \frac{|P \cdot A|}{|A|} = \frac{|P'|}{|A|} \tag{3}$$

where *P'* is the set of real usability problems found by UEM_P and, if *X* is any generic set, $|X|$ is the cardinality of set *X*.

Weighting thoroughness with severity ratings provides a measure that would reveal a UEM's ability to find all problems at all severity levels. Such a measure can be defined by starting with the definition of thoroughness of Equation 2

$$Thoroughness = \frac{\text{number of real problems found}}{\text{number of real problems that exist}} \tag{4}$$

and substituting weighted counts instead of simple counts of problem instances

$$Weighted\ Thoroughness = \frac{s\ (rpf_i)}{s\ (rpe_i)} \tag{5}$$

where $s(u)$ is the severity of an arbitrary usability problem u, rpf_i is the ith real problem found by the UEM in the target system, and rpe_i is the ith real problem that exists in the target system. This kind of measure gives less credit to UEMs finding mostly low-severity problems than ones finding mostly high-severity problems.

However, for many practitioners who want UEMs to find high-severity problems and not even be bothered by low-severity problems, this kind of thoroughness measure does not go far enough in terms of cost effectiveness. For them, perhaps the breakdown of thoroughness at each level of severity is better:

$$Thoroughness(s) = \frac{\text{number of real problems found at severity level (s)}}{\text{number of real problems that exist at severity level (s)}} \tag{6}$$

Practitioners will be most interested in thoroughness for high levels of severity (high values of s) and can ignore thoroughness for low severity. Or a measure of the average severity of problems found by a given UEM, independent of thoroughness, might be more to the point for some practitioners

$$s_{avg}(UEM_A) = \frac{s\ (rpf_i)}{\text{number of real problems found by } UEM_A} \tag{7}$$

and this could be compared to the same measure for other UEMs or to the same measure for the problems existing in the target system:

$$s_{avg}(exist) = \frac{s\ (rpe_i)}{\text{number of real problems that exist}} \tag{8}$$

The previous definition would identify UEMs good at finding the most important problems, even UEMs that do not score the highest in overall thoroughness.

If researchers believe severity is important enough, they can include it in the ultimate and actual criteria as another way to enhance the criterion definition. By including severity, researchers introduce the problem of finding an effective actual criterion that captures "severity-ness" because there is no absolute way to determine the real severity of a given usability problem.

Researchers planning to use severity ratings as part of the criteria for comparing UEMs, however, should be cautious. Nielsen (1994), using Kendall's coefficient of concordance, found interrater reliability of severity ratings so low that individual ratings were shunned in favor of averages of ratings over groups of inspectors.

Nielsen (1994) then used the Spearman–Brown formula for estimating the reliability of the combined judgments and found the group ratings more reliable.

In any discussion of severity, it should be noted that not all users who encounter a problem do so at the same level of severity. Indeed, severity of a usability problem to a user can depend on many factors (e.g., user's background, recent experience, etc.) outside the interaction design and the UEM by which the problem was discovered. A usability problem can block one user's task performance, but another user might recover quickly from the same problem. Thus severity rating is ultimately the responsibility of the usability engineering practitioners, taking into account frequency of problem detection, problem impact over the user community, and likelihood of use of the corresponding part of the interface.

5.2. Validity

In general terms, validity is a measure of how well a method does what it is intended to do. Validity also has a rather nice analogy in the field of information storage and retrieval to the concept of precision, another measure of retrieval performance of an information system from a target document collection (Salton & McGill, 1983). Precision is also based on relevance of a document to a query, analogous to the concept of realness in UEMs. Relevance is the criterion for measuring precision. Precision is the proportion of the documents retrieved by an information system that are relevant:

$$Precision = \frac{\text{number of relevant documents found}}{\text{total number of documents retrieved}} \tag{9}$$

Analogously, Sears (1997) defined *validity* as a measure indicating the proportion of problems found by a UEM that are real usability problems:

$$Validity = \frac{\text{number of real problems found}}{\text{number of issues identified as problems}} \tag{10}$$

Validity and thoroughness can be computed using the same data—usability problem sets generated and the realness criterion. For example, a UEM that found 20 usability problems in a target system, of which only 5 were determined (by criteria yet to be discussed) to be real, would have a validity rating in this case of $5/20 = 0.25$. UEMs with low validity find large numbers of problems that are not relevant or real, obscuring those problems developers should attend to and wasting developer evaluation, reporting, and analysis time and effort.

As mentioned previously, computing validity in terms of sets, we get

$$Validity = \frac{|P \cdot A|}{|P|} = \frac{|P\phi|}{|P|} \tag{11}$$

In Section 4.1, we explored the union of usability problem sets produced by all UEMs being compared as a standard set of existing real usability problems. This technique yields a better thoroughness measure if the number of methods being compared is relatively large, increasing confidence that almost all the real problems have been found by at least one of the methods. However, one negative effect of this approach is to eliminate validity as a metric, an effect we feel is important enough to all but preclude the union of usability problem sets as a viable approach, as we explain next.

Suppose a UEM comparison study were conducted to compare UEM_P, UEM_Q, and UEM_R. Let $P(X)$ be the usability problem set found in interaction design X by UEM_P and so on for UEM_Q and UEM_R. In this approach, the union of the output sets of the UEMs being evaluated is used as the output of a standard method, UEM_A:

$$A(X) = P(X) \cup Q(X) \cup R(X) \tag{12}$$

Thus, even though most of these studies do not say so explicitly, they are using this union as the basis of an actual criterion. The number of usability problems in this union is bounded by the sum of cardinalities of the participating usability problem sets:

$$|A(X)| = |P(X) \cup Q(X) \cup R(X)| \le |P(X)| + |Q(X)| + |R(X)| \tag{13}$$

Unfortunately, problems that might be identified by some other approach to be not real are all included in this union, decreasing validity. However, this approach to an actual criterion, by definition, prevents any possibility of detecting the reduced validity. Applying Equation 11 to design X

$$\textit{Validity of } UEM_p = \frac{|P(X) \cap A(X)|}{|P(X)|} \tag{14}$$

Because $A(X)$ is a union containing $P(X)$, $P(X)$ is a proper subset of $A(X)$ and nothing is removed from $P(X)$ when it is intersected with $A(X)$. Thus

$$\textit{Validity of } UEM_p = \frac{|P(X)|}{|P(X)|} \tag{15}$$

which is identically equal to 1.0.

In other words, this approach guarantees that the intersection of the UEM usability problem set and the standard usability problem set (the union) will always be the UEM usability problem set itself. This means that all usability problems detected by each method are always considered real and validity is 100% for all participating methods!

5.3. Effectiveness

As mentioned in the previous two sections, thoroughness and validity have rather nice analogies to the concepts of recall and precision, metrics for information retrieval performance based on the criterion of relevance of a document to a query. Just as neither precision nor recall alone is sufficient to determine information system retrieval effectiveness, neither thoroughness nor validity alone is sufficient for UEM effectiveness. For example, high thoroughness alone allows for inclusion of problems that are not real, and high validity alone allows real problems to be missed. It is possible to capture the simultaneous effect of UEM thoroughness and validity in a figure of merit that we could call *effectiveness*, defined simply as the product of thoroughness and validity:

$$Effectiveness = \text{Thoroughness} \cdot \text{Validity} \qquad (16)$$

Effectiveness has the same range of values as thoroughness and validity, from 0 to 1. Where either thoroughness or validity is low, effectiveness will be low also. If a UEM achieves a high thoroughness but at the cost of low validity (or vice versa), this kind of effectiveness measure will be reduced, reflecting a more balanced overview of UEM performance.

However, in information retrieval there are often good reasons to prefer an emphasis of either precision or recall over the other measure. For example, users are often focused on finding what they are looking for, even at the cost of having to sort through some irrelevant items retrieved. Manning and Schutze (1999, pp. 269–270) described a weighted combination of precision and recall called the F measure, a variation of van Rijsbergen's (1979, p. 174) E measure

$$F = \frac{1}{\alpha(1/P) + (1-\alpha)(1/R)} \qquad (17)$$

where P is precision, R is recall, and α is a factor that determines the weighting of precision and recall. A value of $\alpha = 0.5$ is often chosen for equal weighting of P and R. With this α value, the F measure simplifies to $2PR/(R + P)$.

Similarly, the goal of UEM usage is to find usability problems, and evaluators are often willing to accept the cost of sorting through a certain number of false positives to achieve a reasonable level of thoroughness. Thus, a weighted measure for UEM performance could be defined as

$$F = \frac{1}{\alpha(1/\text{Validity}) + (1-\alpha)(1/\text{Thoroughness})} \qquad (18)$$

where, again, α is the factor used to set the weighting between thoroughness and validity.

Several authors, such as Gray and Salzman (1998), have alluded to the concepts of hits, misses, false alarms, and correct rejections in the context of UEM outputs, concepts closely related to thoroughness, validity, and effectiveness. These concepts originated with hypothesis testing error types explained in most modern books on statistics or experimental research (for example, Keppel, 1991, or Winer, Brown, & Michels, 1991) and adapted for signal detection theory (Egan, 1975; Swets, 1964). Further adapting this terminology to usability problem detection, we can identify four cases, as shown in Figure 5, to describe the accuracy of a UEM with respect to the realness of the problems it detects, as determined by some actual criterion, A.

High thoroughness is achieved in a UEM by realizing most of the correct hits and by avoiding most of the Type II errors (misses). Similarly, high validity of a UEM derives from avoiding most of the Type I errors (false alarms) and realizing most of the correct rejections. False alarms do not affect thoroughness, but do detract from validity. Misses do not affect validity but do detract from thoroughness. The intersection in the center and the area outside both ovals represents the areas of highest effectiveness, where the UEM is in agreement with the actual criterion.

Gray and Salzman (1998) said that because we do not have access to truth, a diagram such as Figure 5 is misleading. This is true but is exactly why we use actual criteria in experimental design. Figure 5 is only about how performance of a UEM compares with the actual criteria. As discussed in Section 3, the question of truth or faithfulness of the actual criteria to the ultimate criteria is isolated in the part of the process where actual criteria are selected and such questions are out-of-bounds in the discussion of Figure 5. Once established, actual criteria liberate the experimenters from concern about this kind of truth—that is, from the question of how well the output of a UEM matches the truth about real problems existing in a given target system. The point of having actual criteria is to isolate this question of truth to be an issue only between the ultimate criterion and the

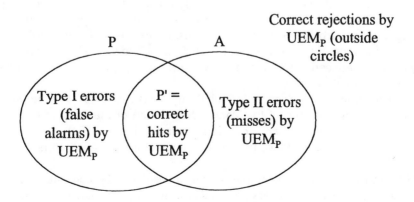

FIGURE 5 Venn diagram of comparison of a UEM$_P$ usability problem set against actual criterion set A.

actual criterion and restrict it to the domain of actual criteria selection, as described in Section 2. We do the best we can at establishing a suitable actual criterion, which then stands instead of the ultimate criterion (truth) and becomes the standard for determining realness of usability problems for the study. During a UEM study, this question of truth (how well a UEM or the actual criterion approximates the ultimate criterion) is not applicable, and the study can be performed without the need to look beyond the actual criterion for truth. The question of truth during a UEM study is, rather, about how well a UEM matches the actual criterion. We do have access to that truth, and that is what is shown in Figure 5. If a better approximation can later be found to the truth of the ultimate criterion, researchers can return to the step of selecting an actual criterion and studies can be repeated.

5.4. Reliability

Reliability of a UEM is a measure of the consistency of usability testing results across different users of the UEMs (evaluators). Usually it is desirable that UEM output be evaluator independent. Pearson's r is an index that describes the extent to which two sets of data are related. This has been used as a measure of reliability in the context of usability problem sets (Nielsen, 1994). From our own experience, we believe agreement is more useful than correlation for this kind of reliability measure.

As a formal measure based on agreement, reliability is an index of agreement between two or more sets of nominal identification, classification, rating, or ranking data. Cohen's (1960) kappa is one example of a reliability measure. *Kappa* is a measure of the proportion of agreement beyond what would be expected on the basis of chance. Kappa has an approximately normal distribution and can be used to test the null hypothesis of no agreement beyond the chance level. Cohen's original kappa ranged from –1 to +1, but negative values for kappa do not correspond to reality in our application—where kappa is, therefore, scaled between 0 and 1, with 0 corresponding to only *chance agreement* and 1 corresponding to *perfect agreement*. Although the original concept of kappa is limited to assessing agreement between two participants, an extension (Fleiss, 1971) permits comparing agreement among several participants. The extension also produces a kappa value between 0 and 1 and allows testing for agreement by reference to the normal distribution.

There are other ways to compute a reliability measure. Sears (1997) measured reliability by using the ratio of the standard deviation of the number of problems found to the average number of problems found. Nielsen (1994) also used Kendall's coefficient of concordance to assess agreement among evaluators making severity ratings.

Although it is usually desirable for UEM results to be consistent or reliable across different individual users, the goals for developing a UEM and the ecological validity of a study for evaluating it will depend on how the UEM is used. Because a UEM typically gives low thoroughness for an individual inspector (e.g., an approximate average of 30%; Nielsen, 1994; Virzi, 1990, 1992; Wright & Monk,

1991), UEMs are usually applied by a group of inspectors and the individual results merged (Nielsen, 1994). If that is how a UEM is used in practice, a realistic comparison study of such methods should be based on group results. Individual results might still be of interest in understanding and tuning the method, but for purposes of comparing UEMs, method performance measures (such as thoroughness and reliability) and method cost for this kind of UEM should be represented by its application within groups rather than by individuals.

This situation also illustrates how UEM developers must approach their goals with care. In most UEMs, low individual reliability means high variability among evaluators, which means that merging results over a group of evaluators will give higher overall thoroughness. The high variability across individual inspectors gives breadth to the union of results.

Although it is reasonable for tool developers to aspire to improve individual inspector reliability by standardizing the inspection process, the standardization can remove the individual variability without improving the individual detection rate, which has the undesired side effect of reducing group thoroughness. If the UEM (particularly inspection method) designers achieve higher individual reliability by narrowing the view of inspectors to some standard guidelines or heuristics and a standard way to apply them, in effect pointing all the inspectors down the same path, it could result in cutting off the broadening effect of individual variation. Thus, it is probably better for UEM developers to strive for higher thoroughness first, and often reliability—at least group reliability—will improve as well in the process.

5.5. Downstream Utility

John and Marks (1997) stand almost alone in their consideration in a UEM study of downstream utility (usefulness in the overall iterative usability engineering process after usability data gathering) of UEM outputs, which depends on the quality of usability problem reporting. We agree with exploring downstream utility and with the quality of usability problem reports as important facets of the overall usability engineering process. John and Marks described evaluating the downstream ability of UEM outputs to suggest effective redesign solutions through usability testing of the redesigned target system interface. This approach has the laudable objective of finding the UEMs that add value or utility in the change process, but inclusion of a more extensive process with high variability brings into question the feasibility of a controlled study of this effect. The iterative cycle of interaction design and redesign is anything but a well-specified and consistent process, depending greatly on team and individual skills, experience, and project constraints. Also, the quality of problem reports is not necessarily an attribute of just the UEM. Many UEMs are designed to detect usability problems, but problem reporting is left to the evaluators using the UEM. In such cases, problem report quality will vary greatly according to the skills of the individual reporter at communicating complete and unambiguous problem reports. Further, usability practitioners do not usually fix all problems found by UEMs. A process leading to fixing the wrong problems, even if with high-quality fixes, might not be most cost effective. In sum, including steps to make the design changes suggested by a UEM and to retest the resulting usability is commendable in UEM studies, but it bears more development.

For UEM studies, we suggest separating (when possible) the treatment of problem reporting and redesign from the basic comparison of UEM performance such as thoroughness and validity, treating UEMs as only functions that produce usability problem lists. The important functions of usability problem classification and reporting and of finding redesign solutions can be treated as separate processes. This approach would give researchers the best chance to match up the best UEM for detecting valid problems with the best techniques for classification, reporting, and connecting to design features.

Researchers who have studied usability problem extraction, description, and classification (Hartson et al., 1999; Lavery, Cockton, & Atkinson, 1997) have made just this kind of separation: regarding these as separate functions of usability engineering support tools for classification and reporting, used in conjunction with UEMs. In fact, most methods for problem description and reporting are independent of the evaluation method and can be used with any UEM.

Perhaps an alternative way to evaluate postusability-testing utility of UEM outputs is by asking real-world usability practitioners to rate their perceptions of usefulness of problem reports in meeting their analysis and redesign needs within the development cycle for the interaction design in their own real development environment. This kind of approach would, of course, have to be validated by the kind of study John and Marks (1997) reported.

5.6. Cost Effectiveness

Just as the ultimate criterion for evaluating a military plane depended on its performance in real battles under real battlefield conditions, we have defined the ultimate criterion for UEM performance in terms of how well it detects or identifies real usability problems in real interaction designs. For many that is the end of the story, but the people who make buying decisions for military planes know there is more. To them, the real criterion for the airplane is to win battles but to do so at the lowest cost in dollars, human lives, and collateral damage. We see the same story in usability practice. For practitioners, the goal is to find real usability problems, to do so with maximum effectiveness, and to do it at the lowest cost possible. To capture this practical consideration, we include cost (e.g., cost to learn and cost to use a UEM) as a metric. Combining cost with our effectiveness metric also yields cost effectiveness, a measure of efficiency.

Good choices for actual criteria would then take efficiency into account. Of course efficiency, in terms of cost and performance, must be defined quantifiably to be compared. As an example, one can combine our effectiveness measure in a quotient with cost to yield cost effectiveness of UEM usage. Cost can be measured as a function of method or tool use, including the fixed overhead of learning a method or tool combined with variable time and effort of applying it. Perhaps the biggest difficulty in getting a measure one can have confidence in is in estimating cost quantitatively and doing it accurately and consistently.

6. REVIEW OF UEM STUDIES

In the spirit of the Gray and Salzman (1998) review of UEM comparison studies, our own research interests led us to explore the relative benefits of various experimental techniques used in UEM comparison studies. We initially approached UEM comparative studies through the use of meta-analysis techniques, attempting to accumulate experimental and correlational results across independent studies. For this effort, we used several criteria to ensure that selected studies were comparable and within the focus of the meta-analysis. The criteria we used were

- The study must involve software usability evaluation.
- A comparison must be made in this as a study of UEMs, using laboratory-based testing with users as a standard of comparison.
- Summary statistics must be reported in the study such that effect sizes can be calculated. Relevant statistics include percentages of problems (of the total) detected by any one UEM (thoroughness) and validity scores.

As soon as we began the meta-analysis process, we realized that a significant majority of the comparison studies in the HCI literature on UEM effectiveness did not provide the descriptive statistics needed to perform a meta-analysis. This confirms the Gray and Salzman (1998) concern with statistical conclusion validity of five popular UEMs in which formal statistical tests were often not included. In addition, many studies did not compare their results to a standard such as laboratory-based testing with users. UEM comparison studies also varied significantly in terms of the criteria used to make comparisons. Criteria included measures that ranged from cost effectiveness to thoroughness, with only a few studies consistently using the same criterion for comparisons. This incompleteness and inconsistency present barriers to meta-analysis, perhaps symptoms of a field that is still very young. In the end, we had to relax our own criteria for selecting studies, giving up on meta-analysis in favor of a descriptive summary of key studies.

We were able to find 18 studies that we could identify as featuring a comparison of UEMs in terms of thoroughness or validity or both. The comparisons were usually among UEMs or among different usage conditions for a single UEM (e.g., applied to different software systems). The 18 studies we identified do not represent the entire population of UEM comparison studies. A full analysis of UEM comparison studies would also embrace such issues as severity, experts versus nonexperts, teams versus individuals, cost, guidelines versus no guidelines, and so forth. However, the 18 studies we selected allowed us to make some conclusions based on issues discussed in earlier sections of this article. A summary comparison of these 18 studies is provided in Table 1, as adapted from Andre, Williges, and Hartson, 1999.

A majority of the UEM comparison studies (14) used the thoroughness measure for comparison. Examining the thoroughness studies in closer detail, we found 7 studies specifically comparing the heuristic evaluation technique with other UEMs. The heuristic evaluation technique was reported as having a higher thor-

Table 1: Summary of UEM Effectiveness Studies

Study	Methods (Subjects)	Thoroughness	Validity	Notes
Bastien & Scapin (1995)	EC (10) NM (10)	EC > NM EC (M = 89.9, SD = 26.2) NM (M = 77.8, SD = 20.7) p < .03		• NM, participants just listed problems without a method guiding them • Study provided M, SD, and p values
Bastien, Scapin, & Leulier (1996)	EC (6) ISO (5) NM (6)	EC > ISO/NM EC (M = 86.2, SD = 12.7) ISO (M = 61.8, SD = 15.8) NM (M = 62.2, SD = 13.8) p <.01		• Study provided M, SD, and p values
Beer, Anodenko, & Sears (1997)	CW (6) TA (6)	TA > CW p < .001		• TA > CW for major, minor, and cosmetic problems
Cuomo & Bowen (1992) Cuomo & Bowen (1994)	HE (2) CW (2) GR (1)	GR > HE > CW	CW > HE > GR CW (58%) HE (46%) GR (22%)	• Not reported: M, SD, and p values • CW: Team approach
Desurvire, Kondziela, & Atwood (1992) Desurvire & Thomas (1993)	HE (3) CW (3) PAVE (3) UT (18)		HE > PAVE > CW HE (44%) PAVE (37%) CW (28%)	• Not reported: SD and p values • PAVE improved DV and NE performance
Doubleday, Ryan, Springett, & Sutcliffe, (1997)	HE (5) UT (20)	HE > UT HE (86) UT (38)		• Not reported: M, SD, and p values • 39% of UT problems not identified by HE • 40% of HE problems not identified by UT
Dutt, Johnson, & Johnson (1994)	HE (3) CW (3)	HE > CW		• Not reported: percentage, M, SD, and p values
Jeffries, Miller, Wharton, & Uyeda (1991)	HE (4) CW (3) GR (3) UT (6)	HE > CW/GR > UT HE (50%) CW (17%) GR (17%) UT (16%)		• Not reported: SD • HE also found highest number of least severe problems • CW and GR essentially used a team of three people, not individuals

Study	Method (N)	Results	Results	Notes
John & Marks (1997)	CA (1) CW (1) GOMS (1) HE (1) UAN (1) SPEC (1)	HE > SPEC > GOMS > CW > CA > CA > UAN HE (31%) SPEC (24%) GOMS (16%) CW (15%) CA (0.08%) UAN (.06%)	CW > SPEC > GOMS > HE > CA/UAN CW (73%) SPEC (39%) GOMS (30%) HE (17%) CA/UAN (0%)	• Not reported: SD and p values • Validity here is the number of problems changed by developer
John & Mashyna (1997)	CW (1) UT (4)		CW (5%)	• Not reported: M, SD, and p values • Case study approach
C.-M. Karat, Campbell, & Fiegel (1992)	IW (6) TW (6) UT (6)	UT > TW > IW $p < .01$		• Not reported: Percentage and SD • Walk-throughs essentially used heuristics for evaluation • Evaluated two different systems, but did not characterize the difference between the two systems
Nielsen & Molich (1990)	HE (various)	HE problems found: 20%–51% (M)		• Not reported: SD and p values • Compared different systems using HE
Nielsen (1990)	TA (36)	TA found 49% (M) of problems		• Not reported: SD and p values
Nielsen (1992)	HE (overall)	HE overall average across six systems was 35%		• Not reported: SD • Nielsen collapsed six HE studies
Sears (1997)	HE (6) CW (7) HW (7)	HE > HW > CW (combining four or five evaluators) Hardware > HE > CW (combining two or three evaluators)	HW > CW > HE	• UT used to determine actual problems • No M or SD reported for thoroughness or validity
Virzi, Sorce, & Herbert (1993)	HE (6) TA (10) UT (10)	HE > TA > UT HE (81%) TA (69%) UT (46%)		• Not reported: SD and p values
Virzi (1990)	TA (20)	TA found 36% (M) of problems		• Not reported: SD and p values
Virzi (1992)	TA (12)	$M = 32\%$, $SD = .14$		• Reported overall detection rate for individuals

Note.   EC = ergonomic criteria; NM = no method; ISO = International Organization for Standards; CW = cognitive walk-through; TA = thinking aloud; HE = heuristic evaluation; GR = guidelines review; PAVE = programmed amplification of valuable experts; DV = Developers; NE = Nonexperts; UT = usability laboratory test; CA = claims analysis; SPEC = reading the specification; GOMS = goals, operators, methods, and selection rules; UAN = user action notation; IW = individual walk-through (essentially used heuristics, not CW process); TW = team walk-through (essentially used heuristics, not CW process); HW = heuristic walkthrough.

oughness rating in 6 out of these 7 studies (85.7%). Thus, a natural conclusion from the thoroughness criterion is that heuristic evaluation appears to find more problems than other UEMs when compared head-to-head, and such a conclusion is often reported in the literature with only a few exceptions. However, many of these studies use a somewhat loose definition of thoroughness based only on a raw count of usability problems found. If these studies had used the tighter definition of thoroughness we have presented in this article, in which only real usability problems count toward the thoroughness measure, thoroughness results might have been somewhat, but probably not greatly, different. Inclusion of the realness criterion in the validity measure would penalize the heuristic method for the relatively high rate of false alarms and low-impact problems it was reported to identify.

On the other hand, use of the validity measure in these studies was disappointing, in general, offering mixed results in terms of identifying a particular UEM that might be more effective for finding problems that impact real users. Many of the studies we reviewed did not explicitly describe how the validity measure was calculated, especially in terms related to a standard such as laboratory-based testing with users.

Because very few studies provided the appropriate descriptive statistics, a robust meta-analysis was nearly impossible. Researchers counting "votes" may be able to conclude realistically that the heuristic evaluation method finds more problems than other UEMs. It would be more profitable to be able to conclude that a particular UEM has the highest effectiveness (per our definition in Section 5.3) and therefore finds just those problems that impact users in real work contexts. Such a conclusion can only be made when the criteria we use to measure effectiveness are relevant to the real work context, highlighting the need for the usability research community to consider carefully the criteria they use in UEM studies.

7. OTHER CONSIDERATIONS IN UEM STUDIES

7.1. UEM Comparison Experiments Should Focus on Qualitative Data Gathering Abilities

As we said in Section 1.4, formative evaluation is evaluation to improve a design and summative evaluation is evaluation to assess a design. UEMs are for gathering qualitative usability data for formative evaluation of interaction design usability. However, as we also mentioned in Section 1.4, some formative UEMs (e.g., laboratory-based usability testing) have a component with a summative flavor in that they also gather quantitative usability data. However, these quantitative data are not intended to provide the statistical significance required in summative evaluation.

The real work of UEMs is to support formative usability data gathering by finding usability problems that can be fixed in an iterative redesign process to achieve an acceptable level of usability. UEM studies should focus solely and explicitly on the ability to gather this qualitative data (e.g., usability problem sets). It is neither feasible nor useful to attempt to compare UEM quantitative data gathering abilities for two main reasons:

- Quantitative data gathered depends on usability goals, such as error avoidance in safety critical systems, walk-up-and-use performance, and ease of learning in public installations, engagement in video games, or long-term expert performance in complex systems. Thus, data gathered can vary independently of the UEM used.
- Quantitative data (e.g., about user task performance) gathering abilities are not usually an inherent part or unique feature of a UEM. Quantitative data represent such variables as time on task, error counts, and user satisfaction scores, all of which are measured in essentially the same way more or less independently of the UEM being used.

Unfortunately, there is confusion about this point. For example, Gray and Salzman (1998) cited, as a drawback of some UEM studies, that "it is not clear that what is being compared across UEMs is their ability to assess usability" (p. 206). Yet in fact, because of the previously mentioned two points, the ability to assess usability is not compared in UEM studies because quantitative data gathering abilities are not comparable. None of the studies Gray and Salzman cited attempted to compare how well UEMs collect quantitative data. In fact, many of the UEMs studied (most inspection methods) do not even produce quantitative user performance data.

Thus, bringing quantitative usability data into discussions of UEM comparisons can add confusion. An example of this potential confusion is seen in this statement by Gray and Salzman (1998):

> When an empirical UEM is used to compare the usability of two different interfaces on some measure(s) of usability (e.g., time to complete a series of tasks) the results are clear and unambiguous: The faster system is the more usable (by that criterion for usability). (p. 206)

First of all, per the previously mentioned points, this statement is out of place in a discussion about experimental validity of UEM studies because it is about an experimental comparison, ostensibly by a UEM being used summatively, of usability between two systems. "Clear and unambiguous" comparison results can come only from a UEM being used for summative evaluation of the interfaces, but the context is about UEMs for formative usability evaluation.

Although qualitative usability data as collected by UEMs are central to UEM studies, readers of Gray and Salzman (1998) could be left with a feeling of ambiguity about their importance. In fact, Gray and Salzman said that "none of the studies we reviewed report systematic ways of relating payoff problems to intrinsic features; all apparently rely on some form of expert judgment" (p. 216). However, that connection of user performance to causes in usability problems is precisely what one gets from qualitative data produced by a UEM. Usability problem lists and critical incidents or verbal protocol data provide relations (attributions of cause) to intrinsic features for observed problems in payoff performance. Evaluators using a UEM may not always get the causes correctly, and they may not always be able to associate a specific cause with a specific shortcoming in performance, but each real usability problem in the output list is potentially the cause of some payoff problem.

7.2. Limitations of UEM Studies

Gray and Salzman (1998) suggested the value of the experimental approach to provide strong tests of causal hypotheses—strong inferences about cause and effect. With proper experimental design for internal validity, one can be sure that a difference in the independent variable (the treatment) is, indeed, the cause of an observed difference in the dependent variable. However, the concern noted by Gray and Salzman regarding the strength of possible inference about causality is very difficult to resolve in the case of UEM studies in which one is comparing one UEM against another that is potentially entirely different. The differences are far too many to tie up in a tidy representation by independent variables, forcing us to compare apples and oranges (J. Karat, 1998).

Monk (1998), in his capsule of the uses and limitations of experimental methods, pointed out that experimental studies are intended for deciding small questions among apples and apples, not grand cross-fruit questions such as deciding what UEM to use and why one is better. When the only independent variable that can be isolated in a treatment is the choice of UEM used, it means a black box view of the UEMs is being used, precluding the possibility of even asking the causality question about why one UEM was better. Causality questions cannot go into more detail than the detail represented by the independent variables themselves.

The black box view of a UEM highlights the fact that the only thing the UEMs can be relied on to have in common is that they produce usability problem lists when applied to a target interaction design (the essential characteristic of a UEM pointed out in Section 1.3). This fact heavily influences the possible choices for UEM performance measures. In essence, it means that an actual criterion to evaluate how well various UEMs produce these lists will have to be based on something revealed in the quality of those lists (e.g., realness) quantified within measures of that usability data.

Although the narrow black box view of UEMs does make comparisons of basic performance measures tractable, it does ignore other aspects and characteristics that could be important distinguishers between UEMs. It is appropriate, for example, to adjust the conditions for UEM application or to adjust the criteria for validity (realness) of usability problems detected. In other words, it is appropriate for UEMs with special talents to compete for attention in a correspondingly tailored arena, as long as (as Gray & Salzman, 1998, pointed out) the ground rules and conditions are the same for each UEM and are made clear in any report of the results. For example, if a given UEM is good at focusing an evaluation instance to narrow the scope of the inspection to just meet evaluation needs at a certain stage of product development, it might represent a cost savings over another UEM that would force broader than necessary inspection at that same juncture in development. This would be a fair, albeit narrow, comparison of apples and apples or at worst apples compared with oranges trying to act like apples—which is acceptable if practitioners do, in fact, need apples today.

7.3. Classification of Usability Problems for Structured Descriptions

Because classification aids description, some researchers are developing usability problem classification schemes to support more uniform problem description. As Gray and Salzman (1998) pointed out, "one would think our discipline would already have a set of common categories for describing our most basic concepts in usability, but no standard categories yet exist." Gray and Salzman encountered three types of classification in the studies they reviewed: categories created in the course of a study to account for data collected, lists of attributes from the literature, and one by Cuomo and Bowen (1994) based on theory. Lavery et al. (1997) and Cockton and Lavery (1999) also found they needed a classification scheme to compare and match problems predicted by analytic methods to problems observed by empirical methods. The Cockton and Lavery framework for structured usability problem extraction is aimed at quality usability problem description and reporting through reliable problem extraction.

Like Cuomo and Bowen (1994), we have followed the path blazed by Norman (1986) with his seven-stages-of-action model of interaction and have developed a detailed classification framework of usability attributes called the *User Action Framework* (Hartson et al., 1999) with which to classify usability problems by type. Often usability problems with underlying similarities can appear very different on the surface and vice versa. The User Action Framework allows usability engineers to normalize usability problem descriptions based on identifying the underlying usability problem types within a structured knowledge base of usability concepts and issues. The User Action Framework provides a highly reliable (Andre, Belz, McCreary, & Hartson, 2000) means for a detailed classification of usability problems by a hierarchical structure of usability attributes, locating a usability problem instance very specifically within the usability or design space. The set of attributes, determined a node at a time along a classification path, represents a kind of "canonical encoding" of each usability problem in standard usability language.

7.4. Usability

We find it interesting that very little of the UEM evaluation literature, if any, mentions UEM usability as a UEM attribute or comparison measure. Perhaps we, as usability researchers and practitioners, should apply our own concepts to our own tool development. Surely UEM usability is a factor in determining its cost to use and, therefore, ought to be part of a criterion for selecting a UEM.

8. CONCLUDING DISCUSSION

Although categories of UEMs are becoming somewhat well defined in the HCI discipline, techniques for evaluating and comparing UEM effectiveness are not yet well established. We believe it is possible to develop stable and consistent criteria for UEM effectiveness. Thoroughness, validity, and reliability appear to form the core of criterion measures researchers should continue to investigate. Thoroughness and validity measures must take into account the question of usability problem realness, and laboratory-based testing with users appears to be an effective way to provide a standard set of real usability problems. Although not an exact replication of real work contexts, user-based laboratory testing can provide a good indication of the types of problems that actually impact users, given a broad enough range of scenarios of use and appropriate participant heterogeneity. Another possibility for researchers is to push for examining problems that real users do encounter in real-world contexts, using field studies and remote usability evaluation (Hartson & Castillo, 1998). Although this kind of usability data is very useful to practitioners, the main difficulty with these methods for researchers is the lack of controls on tasks users perform, leading to an inability to compare results among such UEMs.

We also believe that both usability researchers and usability practitioners will benefit from methods and tools designed to support UEMs by facilitating usability problem classification, analysis, reporting, and documentation, as well as usability problem data management (Hartson et al., 1999). In the context of UEM evaluation, we regard a reliable usability problem classification technique as essential for comparing usability problem descriptions, required at more than one point in UEM studies, although this use of problem classification will probably be more useful to researchers studying UEMs than to practitioners trying to use UEMs to improve systems. We also regard problem classification as very useful for practitioners too, if it can help isolate the cause and possible solutions.

Finally, researchers should consider ways to reduce criterion deficiency and criterion contamination. We believe the easiest way to reduce criterion deficiency is through the use of several measures in the actual criterion, each focusing on a different characteristic of the UEM. In addition, it may be possible to examine how multiple measures can be combined into a composite measure that has a stronger relation to the ultimate criteria.

At this point in the HCI field, it appears to be nearly impossible to do an appropriate meta-comparison of usability studies. We believe there are two reasons that contribute to the challenge of comparing UEMs. First, the field of UEMs is young compared to social science disciplines in which baseline studies are frequently performed. Because of its youth, baseline comparative studies are still almost nonexistent. Second, the methods for usability evaluation themselves are not stable. In fact, they continue to change because human–computer systems, their interaction components, and their evaluation needs change rapidly, requiring new kinds of UEMs and constant improvement and modifications to existing UEMs.

It was our objective in this article to help alleviate the problems of variation, incompleteness, and inconsistency in UEM evaluation and comparison studies. Studies often suffer from the apples and oranges problem, mixing different factors that cannot be compared. We limited the scope to include only UEMs intended for formative usability evaluation. Many UEM studies are based on metrics, without a firm association with comparison criteria. We urged careful attention to comparison criteria—both by researchers who perform UEM evaluation and comparison studies and by practitioners who use the results of those studies—and we explored the considerations for, and the consequences of, actual criteria selection. Many existing studies of

UEMs seem to be done without understanding of (or at least are reported without discussion of) the alternatives for establishing realness of usability problems found by the UEMs being studied. We highlighted the central role that usability problem realness plays in criteria and suggested the advantages and pitfalls of various ways to determine realness of usability problems found by UEMs being studied.

UEM performance measures used in comparison studies are often ill defined and narrow in scope. Established measures are often used without discussion of their meaning in the context of a study, especially their significance to practitioners trying to decide among alternative UEMs. We attempted to broaden and structure understanding of UEM performance measures, to look at what each measure means for the practitioner, and to show how different measures are required for different criteria and different goals for use.

We think of these suggestions and definitions not as the final word, but more as a point of departure for more discussion and collaboration in bringing more science to bear on UEM development, evaluation, and comparison. Finally, we illustrated our points with a brief review of selected UEMs studies.

REFERENCES

Andre, T. S., Belz, S. M., McCreary, F. A., & Hartson, H. R. (2000). Testing a framework for reliable classification of usability problems. In *Human Factors and Ergonomics Society Annual Meeting* (pp. 573–576). Santa Monica, CA: Human Factors and Ergonomics Society.

Andre, T. S., Williges, R. C., & Hartson, H. R. (1999). The effectiveness of usability evaluation methods: Determining the appropriate criteria. In *Human Factors and Ergonomics Society 43rd Annual Meeting* (pp. 1090–1094). Santa Monica, CA: Human Factors and Ergonomics Society.

Bastien, J. M. C., & Scapin, D. L. (1995). Evaluating a user interface with ergonomic criteria. *International Journal of Human–Computer Interaction, 7,* 105–121.

Bastien, J. M. C., Scapin, D. L., & Leulier, C. (1996). Looking for usability problems with the ergonomic criteria and with the ISO 9241–10 dialogue principles. In *CHI Conference on Human Factors in Computing Systems* (pp. 77–78). New York: ACM.

Beer, T., Anodenko, T., & Sears, A. (1997). A pair of techniques for effective interface evaluation: Cognitive walkthroughs and think-aloud evaluations. In *Proceedings of the Human Factors and Ergonomics Society 41st Annual Meeting* (pp. 380–384). Santa Monica, CA: Human Factors and Ergonomics Society.

Bias, R. (1991). Walkthroughs: Efficient collaborative testing. *IEEE Software, 8*(5), 94–95.

Bradford, J. S. (1994). Evaluating high-level design: Synergistic use of inspection and usability methods for evaluating early software designs. In J. Nielsen & R. L. Mack (Eds.), *Usability inspection methods* (pp. 235–253). New York: Wiley.

Card, S. K., Moran, T. P., & Newell, A. (1983). *The psychology of human–computer interaction.* Hillsdale, NJ: Lawrence Erlbaum Associates, Inc.

Carroll, J. M., Singley, M. K., & Rosson, M. B. (1992). Integrating theory development with design evaluation. *Behaviour & Information Technology, 11,* 247–255.

Chin, J. P., Diehl, V. A., & Norman, K. L. (1988). Development of an instrument measuring user satisfaction of the human–computer interface. In *CHI Conference on Human Factors in Computing Systems* (pp. 213–218). New York: ACM.

Cockton, G., & Lavery, D. (1999). A framework for usability problem extraction. In *INTERACT '99* (pp. 344–352). London: IOS Press.

Cohen, J. (1960). A coefficient of agreement for nominal scales. *Educational and Psychological Measurement, 20*, 37–46.

Cuomo, D. L., & Bowen, C. D. (1992). Stages of user activity model as a basis for user-centered interface evaluation. In *Annual Human Factors Society Conference* (pp. 1254–1258). Santa Monica, CA: Human Factors Society.

Cuomo, D. L., & Bowen, C. D. (1994). Understanding usability issues addressed by three user-system interface evaluation techniques. *Interacting With Computers, 6*(1), 86–108.

del Galdo, E. M., Williges, R. C., Williges, B. H., & Wixon, D. R. (1986). An evaluation of critical incidents for software documentation design. In *Thirtieth Annual Human Factors Society Conference* (pp. 19–23). Anaheim, CA: Human Factors Society.

del Galdo, E. M., Williges, R. C., Williges, B. H., & Wixon, D. R. (1987). A critical incident evaluation tool for software documentation. In L. S. Mark, J. S. Warm, & R. L. Huston (Eds.), *Ergonomics and human factors* (pp. 253–258). New York: Springer-Verlag.

Desurvire, H. W., Kondziela, J. M., & Atwood, M. E. (1992). What is gained and lost when using evaluation methods other than empirical testing. In A. Monk, D. Diaper, & M. D. Harrison (Eds.), *People and computers Volume VII* (pp. 89–102). Cambridge, England: Cambridge University Press.

Desurvire, H. W., & Thomas, J. C. (1993). Enhancing the performance of interface evaluators. In *Proceedings of the Human Factors and Ergonomics Society 37th Annual Meeting* (pp. 1132–1136). Seattle, WA: Human Factors and Ergonomics Society.

Doubleday, A., Ryan, M., Springett, M., & Sutcliffe, A. (1997). A comparison of usability techniques for evaluating design. In *Designing interactive Systems (DIS '97) Conference* (pp. 101–110). New York: ACM.

Dutt, A., Johnson, H., & Johnson, P. (1994). Evaluating evaluation methods. In G. Cockton, S. W. Draper, & G. R. S. Weir (Eds.), *People and computers, Volume IX* (pp. 109–121). Cambridge, England: Cambridge University Press.

Egan, J. P. (1975). *Signal detection theory and ROC analysis*. New York: Academic.

Ericsson, A., & Simon, H. (1984). *Protocol analysis: Verbal reports as data*. Cambridge, MA: MIT Press.

Flanagan, J. C. (1954). The critical incident technique. *Psychological Bulletin, 51*, 327–358.

Fleiss, J. L. (1971). Measuring nominal scale agreement among many raters. *Psychological Bulletin, 76*, 378–382.

Gray, W. D., & Salzman, M. C. (1998). Damaged merchandise? A review of experiments that compare usability evaluation methods. *Human–Computer Interaction, 13*, 203–262.

Hartson, H. R., Andre, T. S., Williges, R. C., & van Rens, L. S. (1999). The user action framework: A theory-based foundation for inspection and classification of usability problems. In H.-J. Bullinger & J. Ziegler (Eds.), *Human–computer interaction: Ergonomics and user interfaces, Volume 1* (pp. 1058–1062). Mahwah, NJ: Lawrence Erlbaum Associates, Inc.

Hartson, H. R., & Castillo, J. C. (1998). Remote evaluation for post-deployment usability improvement. In *Advanced Visual Interfaces '98* (pp. 22–29). L'Aquila, Italy: ACM Press.

Hartson, H. R., Castillo, J. C., Kelso, J., Kamler, J., & Neale, W. C. (1996). Remote evaluation: The network as an extension of the usability laboratory. In *CHI Conference on Human Factors in Computing Systems* (pp. 228–235). New York: ACM.

Hix, D., & Hartson, H. R. (1993a). *Developing user interfaces: Ensuring usability through product & process*. New York: Wiley.

Hix, D., & Hartson, H. R. (1993b). Formative evaluation: Ensuring usability in user interfaces. In L. Bass & P. Dewan (Eds.), *Trends in software, Volume 1: User interface software* (pp. 1–30). New York: Wiley.

Jeffries, R., Miller, J. R., Wharton, C., & Uyeda, K. M. (1991). User interface evaluation in the real world: A comparison of four techniques. In *CHI Conference on Human Factors in Computing Systems* (pp. 119–124). New York: ACM.

John, B. E., & Marks, S. J. (1997). Tracking the effectiveness of usability evaluation methods. *Behaviour & Information Technology, 16*, 188–202.

John, B. E., & Mashyna, M. M. (1997). Evaluating a multimedia authoring tool with cognitive walkthrough and think-aloud user studies. *Journal of the American Society of Information Science, 48*, 1004–1022.

Kahn, M. J., & Prail, A. (1994). Formal usability inspections. In J. Nielsen & R. L. Mack (Eds.), *Usability inspection methods* (pp. 141–171). New York: Wiley.

Karat, C.-M., Campbell, R., & Fiegel, T. (1992). Comparison of empirical testing and walk-through methods in user interface evaluation. In *CHI Conference on Human Factors in Computing Systems* (pp. 397–404). New York: ACM.

Karat, J. (1998). The fine art of comparing apples and oranges. *Human–Computer Interaction, 13*, 265–269.

Keppel, G. (1991). *Design and analysis: A researcher's handbook.* Englewood Cliffs, NJ: Prentice Hall.

Kies, J. K., Williges, R. C., & Rosson, M. B. (1998). Coordinating computer-supported cooperative work: A review of research issues and strategies. *Journal of the American Society for Information Science, 49*, 776–779.

Landauer, T. K. (1995). *The trouble with computers: Usefulness, usability, and productivity.* Cambridge, MA: MIT Press.

Lavery, D., Cockton, G., & Atkinson, M. P. (1997). Comparison of evaluation methods using structured usability problem reports. *Behaviour & Information Technology, 16*, 246–266.

Lewis, C., Polson, P., Wharton, C., & Rieman, J. (1990). Testing a walkthrough methodology for theory-based design of walk-up-and-use interfaces. In *CHI Conference on Human Factors in Computing Systems* (pp. 235–242). New York: ACM.

Lewis, J. R. (1994). Sample sizes for usability studies: Additional considerations. *Human Factors, 36*, 368–378.

Lund, A. M. (1998). Damaged merchandise? Comments on shopping at outlet malls. *Human–Computer Interaction, 13*, 276–281.

Manning, C. D., & Schutze, H. (1999). *Foundations of statistical natural language processing.* Cambridge, MA: MIT Press.

Marchetti, R. (1994). Using usability inspections to find usability problems early in the lifecycle. In *Pacific Northwest Software Quality Conference* (pp. 1–19). Palo Alto, CA: Hewlett Packard.

Meister, D., Andre, T. S., & Aretz, A. J. (1997). System analysis. In T. S. Andre & A. W. Schopper (Eds.), *Human factors engineering in system design* (pp. 21–55). Dayton, OH: Crew System Ergonomics Information Analysis Center.

Monk, A. F. (1998). Experiments are for small questions, not large ones like "What usability evaluation method should I use?" *Human–Computer Interaction, 13*, 296–303.

Neale, D. C., Dunlap, R., Isenhour, P., & Carroll, J. M. (2000). Collaborative critical incident development. In *Human Factors and Ergonomics Society 43rd Annual Meeting* (pp. 598–601). Santa Monica, CA: Human Factors and Ergonomics Society.

Newman, W. M. (1998). On simulation, measurement, and piecewise usability evaluation. *Human–Computer Interaction, 13*, 316–323.

Nielsen, J. (1989). Usability engineering at a discount. In G. Salvendy & M. J. Smith (Eds.), *Designing and using human–computer interfaces and knowledge-based systems* (pp. 394–401). Amsterdam: Elsevier.

Nielsen, J. (1990). Evaluating the thinking aloud technique for use by computer scientists. In H. R. Hartson & D. Hix (Eds.), *Advances in human–computer interaction* (pp. 69–82). Norwood, NJ: Ablex.

Nielsen, J. (1992). Finding usability problems through heuristic evaluation. In *CHI Conference on Human Factors in Computing Systems* (pp. 373–380). New York: ACM.

Nielsen, J. (1994). Heuristic evaluation. In J. Nielsen & R. L. Mack (Eds.), *Usability inspection methods* (pp. 25–62). New York: Wiley.

Nielsen, J., & Mack, R. L. (Eds.). (1994). *Usability inspection methods.* New York: Wiley.

Nielsen, J., & Molich, R. (1990). Heuristic evaluation of user interfaces. In *CHI Conference on Human Factors in Computing Systems* (pp. 249–256). New York: ACM.

Norman, D. A. (1986). Cognitive engineering. In D. A. Norman & S. W. Draper (Eds.), *User centered system design* (pp. 31–61). Hillsdale, NJ: Lawrence Erlbaum Associates, Inc.

Olson, G. M., & Moran, T. P. (1998). Commentary on "Damaged merchandise?" *Human–Computer Interaction, 13,* 263–323.

Rubin, J. (1994). *Handbook of usability testing.* New York: Wiley.

Salton, G., & McGill, M. J. (1983). *Introduction to modern information retrieval.* New York: McGraw-Hill.

Scriven, M. (1967). The methodology of evaluation. In R. Tyler, R. Gagne, & M. Scriven (Eds.), *Perspectives of curriculum evaluation* (pp. 39–83). Chicago: Rand McNally.

Sears, A. (1997). Heuristic walkthroughs: Finding the problems without the noise. *International Journal of Human–Computer Interaction, 9,* 213–234.

Smith, S. L., & Mosier, J. N. (1986). *Guidelines for designing user interface software* (Technical Report No. ESD– TR–86–278/MTR 10090). Bedford, MA: MITRE Corporation.

Swets, J. A. (1964). *Signal detection and recognition by human observers.* New York: Wiley.

Thompson, J. A., & Williges, R. C. (2000). Web-based collection of critical incidents during remote usability evaluation. In *IEA 2000/HFES 2000 Congress* (pp. 602–605). Santa Monica, CA: Human Factors and Ergonomics Society.

van Rijsbergen, C. J. (1979). *Information retrieval* (2nd ed.). London: Butterworths.

Virzi, R. A. (1990). Streamlining the design process: Running fewer subjects. In *Human Factors and Ergonomics Society 34th Annual Meeting* (pp. 291–294). Santa Monica, CA: Human Factors and Ergonomics Society.

Virzi, R. A. (1992). Refining the test phase of usability evaluation: How many subjects is enough? *Human Factors, 34,* 457–468.

Virzi, R. A., Sorce, J., & Herbert, L. B. (1993). A comparison of three usability evaluation methods: Heuristic, think-aloud, and performance testing. In *Proceedings of the Human Factors and Ergonomics Society 36th Annual Meeting* (pp. 309–313). Seattle, WA: Human Factors and Ergonomics Society.

Wharton, C., Bradford, J., Jeffries, R., & Franzke, M. (1992). Applying cognitive walkthroughs to more complex user interfaces: Experiences, issues, and recommendations. In *CHI Conference on Human Factors in Computing Systems* (pp. 381–388). New York: ACM.

Winer, B. J., Brown, D. R., & Michels, K. M. (1991). *Statistical principles in experimental design.* New York: McGraw-Hill.

Wright, P., & Monk, A. (1991). A cost-effective evaluation method for use by designers. *International Journal of Man–Machine Studies, 35,* 891–912.

INTERNATIONAL JOURNAL OF HUMAN–COMPUTER INTERACTION, *13*(4), 411–419
Copyright © 2001, Lawrence Erlbaum Associates, Inc.

Task-Selection Bias: A Case for User-Defined Tasks

Richard E. Cordes
IBM Corporation

Usability evaluations typically occur throughout the life cycle of a product. A number of decisions and practical biases concerning the tasks selected for usability evaluations can influence the results. A pervasive bias is to select only tasks that are possible to perform with the product under evaluation, introducing a subtle bias for the participants. One way to avoid this problem is to employ user-defined tasks (UDTs) in usability evaluations. In addition, having participants define tasks to perform in a product evaluation allows a more accurate assessment of product usability. This is because UDTs based on users' requirements and expectations should be relatively independent of the functional capabilities of a product. However, there are a number of methodological and practical issues that result from the introduction of UDTs in a usability evaluation. The best approach is to design hybrid evaluations using both UDTs and product-supported tasks.

1. INTRODUCTION

Within the computer industry, a common practice is to have products, before their public release, go through a usability evaluation (Nielsen, 1993). A usability evaluation typically consists of having representative users of a product perform typical product tasks within a controlled laboratory environment (Rubin, 1994). The fundamental goal of the evaluation is to identify usability problems with a product so developers can improve the product before real users encounter these problems (Dumas & Redish, 1993). Sometimes, developers reach conclusions about a product's usability by determining whether a product can meet some predefined usability criteria, for example, successful completion of 98% of the tasks. If the participants in the usability evaluation achieve or exceed the criteria, have a positive impression of the product, and no major usability problems remain in the product, then the typical assumption is that actual users will not have serious usability problems with the product.

Requests for reprints should be sent to Richard E. Cordes, IBM Corporation, 4205 South Miami Boulevard, CX4A/503, Research Triangle Park, NC 27709. E-mail: cordes@us.ibm.com

However, as shown by Molich et al. (1998), there is very little agreement among usability laboratories on the number and nature of problems uncovered by usability testing that follows this methodology. Part of the disagreement involves selecting the proper mix of tasks that is representative of the tasks the intended users will perform with the product. The degree to which usability practitioners achieve this goal can help determine how well they can consistently apply and generalize their results to real-world perceptions of usability (and see more consistent results between laboratories). Although there are numerous biases that can affect the results of usability evaluations (e.g., Cordes, 1992), task selection is a major and often overlooked one. In this article, I focus on task-selection biases, including ways to select tasks that better match users' expectations.

2. TASK-SELECTION BIAS

There are a number of potential biases that can occur when choosing which product tasks to evaluate. Tasks selected for evaluation should be representative of what users will do with the product and must be manageable and suitable for a laboratory evaluation. Such tasks are typically:

- Tasks that are short and fit within a test session. Long tasks reduce the number of participants for a given test schedule. There are also problems of participant availability and attrition for sessions that take longer than 8 hr.
- Tasks the evaluator knows how to do. Some product tasks are subtle, quite complex, and require a higher level of expertise than the practitioner has. For obvious reasons, practitioners do not evaluate tasks that they themselves do not know how to perform. The tasks they select typically represent the domain of product tasks with which they are familiar and know how to do. These are not necessarily the key tasks from the user's perspective.
- Tasks that are consistent with a laboratory environment. Some important tasks (e.g., product migration from a competitor's product) are difficult to replicate in a laboratory environment. Some tasks might be so user specific that it is not possible to construct typical scenarios. Also, there may not be adequate resources available to replicate a complex multiproduct customer environment.
- Tasks the evaluator finds interesting. The evaluator may have a predisposition to focus on tasks addressing product areas that were important or controversial during design or may simply feel more comfortable testing familiar areas of the user interface such as a graphical user interface and pop-up dialogs instead of documentation and messages.
- Tasks with available participants. The availability of participants will affect the makeup of the tasks chosen to evaluate. For example, tasks that require participants from the international community are less likely to undergo evaluation than tasks requiring local participants.

3. "I KNOW IT CAN BE DONE OR YOU WOULDN'T HAVE ASKED ME TO DO IT" BIAS

3.1. Background

The previous list of task-selection biases stems primarily from the practical logistics of conducting usability evaluations. Some are apparent and, once acknowledged, are easy to fix. Others equally apparent are more difficult or impossible to fix (so practitioners must simply accept them). There is, however, another task-selection bias that is less obvious and has far-reaching implications in how practitioners conduct and interpret usability evaluations. In most usability tests, practitioners only select tasks that the product supports. They do not ask participants to perform domain-relevant tasks that the product does not support. If participants know this, it can have a profound effect on their performance and attitude about a product. Part of the implicit demand characteristics (Orne, 1962) of usability studies is that all tasks that participants perform with a product are possible to do with that product. In contrast, users interacting with a new product are learning about a product's capabilities and limitations. Indeed, determining what a product can and cannot do is a fundamental aspect of learning how to use a product. Therefore, in a non-laboratory situation, users may not be so sure that they can do the tasks they want to do with a given product. If users are not certain that they can perform their tasks with a product, this belief can bias the amount of time they are willing to spend learning to use a product to perform a specific task. Consequently, they might be much more likely to give up in times of difficulty and to feel that the product is much more difficult to use.

3.2. Magnitude of the Effect

Cordes (1989) evaluated whether a simple addition to task instructions, one that questions the ability of a product to support all tasks, would have an effect on users' thresholds for giving up when learning to use a product. In the experiment, I investigated whether this change in task instruction would affect the number of "I give up" phone calls made by the participants. In the double-blind study, two groups of 8 people participated in a usability evaluation of a software product under development. The participants received random assignment to one of two groups: control or experimental. Both groups received identical task instructions, except that the experimental group heard on their private videotaped instructions, "As in the real world, don't assume that the product can perform each task that we are going to ask you to do." The control group received identical instructions without this statement. The objective measure of a person giving up on a task was a telephone call to the experimenter.

An analysis of variance was conducted to evaluate the effects of Group, Task, and the Group × Task interaction using the dependent measure of the number of "I give up" phone calls. The two main effects and their interaction were statistically

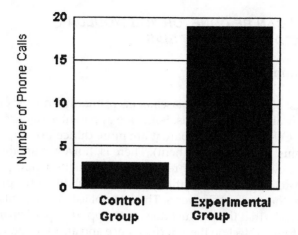

FIGURE 1   Number of "I give up" phone calls for each group.

significant, $F(1, 14) = 7.56$, $p < .016$ (Group); $F(10, 113) = 5.02$, $p < .001$ (Task); and $F(10, 113) = 2.86$, $p < .003$ (Group × Task interaction). The experimental group made a total of 19 phone calls (2.37 per person) compared to 3 (0.37 per person) for the control group (see Figure 1). This is an increase of over 6 times more phone calls due to the instructional change. The significant Group × Task interaction (see Figure 2) indicated that the difference in phone calls between the two groups was related to the task performed. The control group did not give up on any task that the experimental group did not also give up on. However, besides having a higher rate of giving up on these tasks, the experimental group gave up on two additional tasks. This suggests that the experimental group had an overall lower threshold for

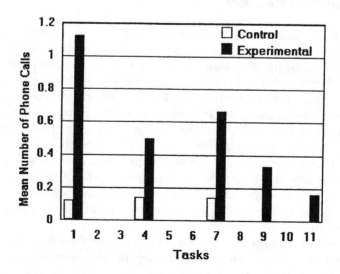

FIGURE 2   Mean number of phone calls by group and task.

giving up. The difference between the groups was attributed to what the participants judged to be the more difficult tasks. As would be expected, tasks judged easy were not sensitive to this manipulation (i.e., the participants gave up only on tasks they perceived to be difficult). Looking at only the "gave up" tasks, the participants' task ratings (using magnitude estimation; Cordes, 1984) showed the experimental group rated these tasks to be over 14 times less difficult than the control group. In addition, the experimental group took a little more than half the time to give up on these tasks as compared to the control group (33 min vs. 59 min). This suggests that not only were the participants in the experimental group willing to give up more frequently, but they also were willing to do so sooner while experiencing less difficulty than the control group.

4. USER-DEFINED TASKS (UDTs)

4.1. Rationale

Cordes's (1989) results demonstrate that task instructions that bring into doubt one's ability to use a product to perform assigned tasks can directly affect a person's threshold for giving up. Therefore, usability evaluations that do not control for this effect have a strong bias in favor of a product achieving a higher successful task completion rate. More people would fail to complete tasks in usability studies if they were less certain that the tasks they attempted were possible to do. Although a change in task instructions appeared to successfully reduce this certainty, this manipulation was artificial. After all, how many products come with a warning label stating "Don't assume you can do everything you want with this product"? Also, hearing an instruction of this nature is unusual and may serve to arouse suspicions and cause an overreaction. For example, participants may believe "Some of these tasks must not be doable with this product or I wouldn't have been given this instruction." Therefore a better method for controlling this bias needs to be explored. Typically, the lack of certainty that a product can perform users' tasks arises from an incomplete match between user requirements and product functions. The Venn diagram in Figure 3 summarizes the situation.

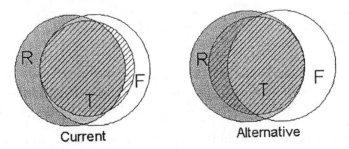

Current   **Alternative**

FIGURE 3   Venn diagram of product requirements (R), product function (F), and human factors testing (T).

Figure 3 shows user requirements (R) and product functions (F). The overlap between these areas represents the degree to which a product has met the user requirements. Typically, usability testing (T in the figure) rests entirely in the F domain, based on tasks that are mostly a subset of R. The usability that users experience, however, depends on the R rather than the F domain. Because users are never completely sure that a product will meet their requirements, they are very likely to try to perform tasks that the product cannot do. Alternatively, if the evaluation included tasks sampled from the R domain, the results would provide a more accurate picture of product usability than sampling strictly from the F domain.

Of course, if product tasks better matched user requirements, then selecting tasks from the F domain would be less of a problem. However, the functions and features that end up in a product represent a compromise between users' requirements and what developers can achieve in a given time, resulting in a reduced set of product functions.

When selecting only the tasks that products can support, practitioners are not evaluating usability from the user's perspective but are only evaluating usability from the perspective of the product functions. When users spend hours trying to perform a task with a product that cannot do it, this experience must have an impact on their perception of product usability. This "learning the capabilities" of a product and how they match user needs is an important component of usability that rarely receives evaluation in laboratory-based usability studies. Field studies that focus on users performing tasks in their real environment do tap into the users' perspective but are typically time consuming, expensive, logistically difficult, and, therefore, less often performed. One way to design laboratory-based usability evaluations that better reflect users' perception of usability and control for the "I know it can be done or you wouldn't have asked me to do it" bias is to include UDTs.

4.2. Definition

UDTs are simply tasks that participants bring into the usability evaluations as opposed to the product-supported tasks (PSTs) that make up most usability evaluations. By having users bring into the laboratory tasks they want to perform and believe they should be able to perform with a product, a truer picture of usability can emerge.

From experimental and practical points of view, however, UDTs can present problems:

1. Because the choice of tasks is up to the participant, practitioners lose control over the content and, potentially, the duration of the evaluation. This can be a problem if participants will perform multiple UDTs, particularly if any of them are not possible to do.

2. The quality of tasks that users bring into the evaluation can vary considerably. For example, in an attempt to test the capabilities of a product, some participants may choose tasks that are currently a challenge for them to perform, but others may choose routine tasks to make sure they are still being supported.

3. Experience with prior or similar versions of the product may influence the tasks the participants will bring into the evaluation. Their expectations may have previously been set, and this will affect their choice of tasks to perform. For example, participants may have a requirement for spell checking on a simple text-editor product, but because the version of this product they are familiar with does not have this feature, the tasks they choose will not include it either.

4. You may not know who the participants will be until they show up, and unless you know this, you cannot collect the tasks needed for the study. Also, if participants are being reimbursed, it is not apparent how to reimburse them for the time spent working on a task to provide for the study.

5. The tasks chosen by the participants will be unique. This introduces the possibility that the tasks chosen by each participant will only be performed by that participant, making it difficult to provide summary analyses (such as means).

Setting time limits for accomplishing tasks can give the practitioner control over the duration of an evaluation. Of course, this control will be at the expense of knowing whether the participants who exceeded the time limit would have eventually given up if they had more time. However, as shown earlier, when task "doability" was uncertain, participants who gave up did so in about half the time, so this issue may not be faced that often. Regarding the second point, although UDTs may vary in quality, it is better to accept this range than to screen the tasks that are submitted by the participants because these tasks come from the user requirements domain. Restricting or selecting the UDTs based on quality implies a more complete knowledge of user requirements than is typically the case. As mentioned previously, based on previous experience, the participants themselves might restrict or otherwise alter their choice of tasks. However, if participants do self-limit their choice of tasks, or even choose tasks that have been dropped from the product, then this is really not a bias problem at all because it accurately reflects their product requirements view. Regarding the difficulty in obtaining UDTs from unknown participants (reimbursement, etc.), a solution would be to specifically hire participants to create the UDTs prior to starting the evaluation then use these UDTs in the study. Overcoming the last problem requires that a study employing UDTs have each participant (or at least more than one participant) perform the same task. A possible solution is to collect a single UDT from each participant before the evaluation and then have each participant perform all of the collected UDTs. For example, if there were 8 participants planned for a study, the evaluator could collect one UDT from each participant and then have each participant perform all eight UDTs. That would permit the use of basic descriptive statistics when summarizing the results. However, this approach also presents some problems:

1. Equating the number of UDTs to the number of participants in a study will only be practical for small numbers of evaluations if the individual tasks take long to complete. For example, it may be too time consuming to run a study with a larger number of participants if each task takes on average longer than one-half hr.

2. For each participant, one of the tasks to perform will be the task that the participant brought to the evaluation. It is likely that the participant will perceive this task

differently than the others (e.g., the participant may have a greater interest in or more motivation to perform this task), introducing some extraneous variability into the study.

A way to solve the first problem is to randomly select a manageable number of tasks from the group of participants. For example, if 12 tasks solicited from 12 participants would incur too long of a session, then the evaluator might randomly select three of the UDTs for inclusion in the study. Regarding the second problem, this might not be a liability because in the real world users will perform some tasks of their own choosing and others that they receive an assignment to perform. A bigger concern is that in the time interval between submitting the task to the practitioner and actually performing it, a participant may plan the effort and, in some cases, perform the task ahead of time. One way to address this problem is for the practitioner to collect tasks from at least t more participants than he or she plans to observe ($n^* = n + t$). The practitioner then randomly selects the t tasks and excludes from the study the t participants who provided these tasks. For example, in a study with only time for 12 participants performing three tasks, you collect tasks from 15 participants, randomly select three tasks, and then exclude the 3 participants who provided the tasks.

Another way to address the second problem is to make the a priori decision not to collect data from participants on the task that they contributed to the study and treat these data points as missing values.

Regardless of how UDTs are employed, their introduction can only serve as a method for controlling the "I know it can be done or you wouldn't have asked me to do it" bias if the participants know that at least some of the tasks came from other participants and that the tasks were not selected or tested for doability by the practitioner. Rather than identifying the origin of each task, this information can best be conveyed by simply telling them that these tasks come from their fellow participants and that they are all going through them for the first time. This is an honest and realistic statement that implicitly brings into question the doability of all the tasks without sounding artificial or manipulative.

4.3. Recommendations and Discussion

Although practitioners should adopt UDTs in their usability evaluations, they should not stop using PSTs. PSTs serve a very useful purpose by

1. Assuring the evaluation of most of the functions of a product.
2. Allowing the evaluator to include tasks that users might perform rarely and, therefore, would be unlikely to be UDTs.
3. Evaluating tasks that are not currently part of the user requirements.
4. Enabling the experimenter to focus on specific tasks suspected to include usability problems.

Usability practitioners should design usability studies that incorporate UDTs with their PSTs. This hybrid approach allows a more realistic assessment of product usability through the introduction of tasks that are based on user requirements and are independent of product capabilities. This approach can also help practitioners avoid the "I know it can be done or you wouldn't have asked me to do it" task-selection bias by implicitly bringing into doubt the doability of all the tasks that participants are asked to perform. By incorporating UDTs into usability evaluations, practitioners will be in a position to assess product usability from the users' perspective with greater accuracy.

Although my colleagues and I have successfully followed the recommendations presented in this article, more systematic work needs to be done to validate the benefit of UDTs and their effect on the "I know it can be done or you wouldn't have asked me to do it" bias. No study comparable to the one highlighted in this article has been performed to evaluate the effectiveness of adopting UDTs in controlling this bias. For example, what effect do actual impossible tasks have on this bias? Participants' willingness to give up on a task might be quite different if they find that all prior tasks were doable versus only 10%. Also, is there an optimum mix of UDTs and PDTs that produces the best results in terms of incorporating user requirements and evaluating product functionality? Perhaps one UDT per participant is sufficient to be able to generalize to real-world product usage, but more likely more are necessary. Finally, if we hope to improve the practice of usability, additional studies need to be performed to investigate the ramifications of the task-selection biases and remedies discussed in this article.

REFERENCES

Cordes, R. E. (1984). Application of magnitude estimation for evaluating software ease-of-use. In G. Salvendy (Ed.), *First USA–Japan Conference on Human Computer Interaction* (pp. 199–202). Amsterdam: Elsevier.

Cordes, R. E. (1989). Are software usability tests biased in favor of your product? *Proceedings of the 28th ACM Annual Technical Symposium* (pp. 1–4). Gaithersburg, MD: ACM.

Cordes, R. E. (1992). Bias in usability studies: Is this stuff science? *HFS Test and Evaluation Newsletter, 7*(2), 2–5.

Dumas, J. S., & Redish, J. C. (1993). *A practical guide to usability testing.* Norwood, NJ: Ablex.

Molich, R., Bevan, N., Butler, S., Curson, I., Kindlund, E., Kirakowski, J., & Miller, D. (1998). Comparative evaluation of usability test. In *Proceedings of the Usability Professionals Association 1998 Conference* (pp. 189–200). Washington, DC: UPA.

Nielsen, J. (1993). *Usability engineering.* Boston: Academic.

Orne, M. T. (1962). On the social psychology of the psychological experiment: With particular reference to demand characteristics and their implications. *American Psychologist, 17,* 776–783.

Rubin, J. (1994). *Handbook of usability testing: How to plan, design, and conduct effective tests.* New York: Wiley.

INTERNATIONAL JOURNAL OF HUMAN–COMPUTER INTERACTION, 13(4), 421–443
Copyright © 2001, Lawrence Erlbaum Associates, Inc.

The Evaluator Effect: A Chilling Fact About Usability Evaluation Methods

Morten Hertzum
Centre for Human–Machine Interaction
Risø National Laboratory, Denmark

Niels Ebbe Jacobsen
Nokia Mobile Phones, Denmark

Computer professionals have a need for robust, easy-to-use usability evaluation methods (UEMs) to help them systematically improve the usability of computer artifacts. However, cognitive walkthrough (CW), heuristic evaluation (HE), and thinking- aloud study (TA)—3 of the most widely used UEMs—suffer from a substantial evaluator effect in that multiple evaluators evaluating the same interface with the same UEM detect markedly different sets of problems. A review of 11 studies of these 3 UEMs reveals that the evaluator effect exists for both novice and experienced evaluators, for both cosmetic and severe problems, for both problem detection and severity assessment, and for evaluations of both simple and complex systems. The average agreement between any 2 evaluators who have evaluated the same system using the same UEM ranges from 5% to 65%, and no 1 of the 3 UEMs is consistently better than the others. Although evaluator effects of this magnitude may not be surprising for a UEM as informal as HE, it is certainly notable that a substantial evaluator effect persists for evaluators who apply the strict procedure of CW or observe users thinking out loud. Hence, it is highly questionable to use a TA with 1 evaluator as an authoritative statement about what problems an interface contains. Generally, the application of the UEMs is characterized by (a) vague goal analyses leading to variability in the task scenarios, (b) vague evaluation procedures leading to anchoring, or (c) vague problem criteria leading to anything being accepted as a usability problem, or all of these. The simplest way of coping with the evaluator effect, which cannot be completely eliminated, is to involve multiple evaluators in usability evaluations.

Morten Hertzum was supported by a grant from the Danish National Research Foundation. We thank Iain Connell for providing us with additional data from Connell and Hammond (1999), Hilary Johnson for access to the data set from Dutt, Johnson, and Johnson (1994), and Rolf Molich for making the data set from Molich et al. (1999) available on the Web (http://www.dialogdesign.dk/cue.html). For insightful comments on earlier versions of this article, we thank Iain Connell, James Lewis, Rolf Molich, and John Rieman.

Requests for reprints should be sent to Morten Hertzum, Centre for Human–Machine Interaction, Systems Analysis Department, Risø National Laboratory, P.O. Box 49, DK–4000 Roskilde, Denmark. E-mail: morten.hertzum@risoe.dk

1. INTRODUCTION

Computer professionals need robust, easy-to-use usability evaluation methods (UEMs). This study is about three prominent UEMs: cognitive walkthrough (CW), heuristic evaluation (HE), and thinking-aloud study (TA). CW was introduced by C. Lewis, Polson, Wharton, and Rieman (1990) and consists of a step-by-step procedure for evaluating the action sequences required to solve tasks with a system. HE was introduced by Nielsen and Molich (1990) and is an informal inspection technique with which evaluators test the system against a small number of interface heuristics. TA was introduced in systems development around 1980 (C. Lewis, 1982) and is probably the single, most important method for practical evaluation of user interfaces (Nielsen, 1993). These three UEMs span large differences in their approach to usability evaluation but share the common goal of supporting systems developers or usability specialists in identifying the parts of a system that cause users trouble, slow them down, or fit badly with their preferred ways of working—commonly termed *usability problems.*

This study is about whether evaluators who evaluate the same system with the same UEM detect—roughly—the same problems in the system. This issue is frequently neglected in UEM research as well as in practice, probably due to lack of awareness of the magnitude of the disagreements combined with prioritizing swift and useful results over reliability and completeness. Several studies have provided evidence that evaluators using the same UEM detect markedly different sets of usability problems when they evaluate a system. Evaluators also seem to differ substantially in their rating of the severity of the detected problems. In this article, we term differences in evaluators' problem detection and severity ratings the *evaluator effect.* We are well aware that a low evaluator effect is only one desirable property of a UEM. We specifically emphasize the distinction between reliability (i.e., the extent to which independent evaluations produce the same result) and validity (i.e., the extent to which the problems detected during an evaluation are also those that show up during real-world use of the system). The evaluator effect is a measure of reliability only. To our knowledge, the validity of UEMs has not been investigated.

In this study, we bring together the results of previous studies and newly derived results from studies that contained the data necessary to investigate the evaluator effect but did not address this issue. With these data, we show that the evaluator effect cannot be dismissed as a chance incident, an artifact of the peculiarities of a single study, or a weakness of a particular UEM. Notably, the evaluator effect is also of concern to TA, which is generally considered the most authoritative method for identifying usability problems. By looking at differences and commonalities of the reviewed studies and UEMs, we then discuss where the UEMs fall short of providing evaluators with the guidance necessary to perform reliable usability evaluations. This leads to input for improving current UEMs but also to the realization that the evaluator effect will, to a considerable extent, have to be managed rather than eliminated.

In the next section, we provide a brief introduction to the three UEMs. In Section 3, two measures of the evaluator effect are defined and discussed. In Section 4, we review 11 UEM studies that provided empirical data on the evaluator effect. As examples, three of these studies are described in more detail. In Section 5, we discuss possible causes for the evaluator effect. Finally, in the concluding section, we aim at advising practitioners on how to cope with the evaluator effect in UEMs.

2. BRIEF INTRODUCTION TO THE METHODS

The following descriptions of CW, HE, and TA are mere introductions provided to give a flavor of how usability evaluation is approached with the three methods. Guidance on how to conduct evaluations with CW, HE, and TA can be found in Wharton, Rieman, Lewis, and Polson (1994), Nielsen (1994a), and Dumas and Redish (1993), respectively.

2.1. Cognitive Walkthrough (CW)

CW (C. Lewis et al., 1990; C. Lewis & Wharton, 1997; Polson, Lewis, Rieman, & Wharton, 1992; Wharton, Bradford, Jeffries, & Franzke, 1992; Wharton et al., 1994) was devised to enable computer professionals to detect usability problems in a user interface based on a detailed specification document, screen mock-ups, or a running system. CW is particularly suited to evaluate designs before testing with users becomes feasible and as a supplement to user testing in situations in which users are difficult or expensive to recruit. Also, CW was initially developed for evaluating walk-up-and-use systems, although it was later applied to more complex interfaces. It has been recommended that CW should be performed by groups of cooperating evaluators, but the descriptions of the method maintain that it can also be performed by evaluators working individually. CW is based on a cognitive theory of exploratory learning called CE+ (Polson & Lewis, 1990; Polson et al., 1992), and a basic understanding of this theory is a definite advantage when performing a walkthrough.

The procedure for CW consists of a preparation phase and an execution phase. In the preparation phase, the evaluator describes a typical user, chooses the tasks to be evaluated, and constructs a correct action sequence for each task. When this is done, the execution phase can begin. For each action in the action sequences the evaluator asks four questions[1]: (a) Will the user try to achieve the right effect?, (b) Will the user notice that the correct action is available?, (c) Will the user associate the correct action with the effect trying to be achieved?, and (d) If the correct action is performed, will the user see that progress is being made toward solution of the task? With the description of the user in mind, the evaluator decides whether each question leads to success or failure. In case of failure, a usability problem has been

---

[1]In the earlier versions of CW, the execution phase consisted of many more questions. However, the many questions made the walkthrough process inordinately tedious and time consuming. In recognition of this, the execution phase of the latest version, described in Wharton et al. (1994) and C. Lewis and Wharton (1997), consists of only four questions.

detected. After all actions have been evaluated, the CW is completed by merging the detected problems into one nonduplicate list.

2.2. Heuristic Evaluation (HE)

HE (Nielsen, 1992, 1993, 1994a; Nielsen & Molich, 1990) is an informal UEM that enables evaluators to detect usability problems in an interface based on screen mock-ups or a running system. The informality has implications for the reliability and coverage of HEs but is considered necessary to get computer professionals to adopt the method. Any computer professional should be able to apply HE, but the informality of the method leaves much to the evaluator. Consequently, the evaluator's skills and expertise have a large bearing on the results. A single, inexperienced evaluator is unlikely to produce sufficiently good results. For this reason, HE prescribes that a small group of evaluators individually inspect the system. In addition, Nielsen (1992) found that the effectiveness of HE can be substantially improved by having usability specialists as evaluators.

The procedure for HE involves having a small group of evaluators examine an interface and judge its compliance with a small set of recognized usability principles—the heuristics. Nielsen (1994a) provided a set of 10 general heuristics, which state that the system should[2] (a) provide visibility of system status; (b) ensure a match between the system and the real world; (c) allow for user control and freedom; (d) be consistent and follow standards; (e) prevent errors; (f) utilize recognition rather than recall; (g) allow for flexibility and efficiency of use; (h) provide aesthetic and minimalist design; (i) help users recognize, diagnose, and recover from errors; and (j) provide help and documentation. Each evaluator goes through the interface and inspects the various dialogue elements and compares them with the heuristics. In addition to the checklist of general heuristics to be considered for all dialogue elements, the evaluator may also consider any additional usability principles or results that seem to be relevant for any specific interface element. To ensure independent and unbiased evaluations, the evaluators are only allowed to communicate and aggregate the results of their evaluations after they have completed their own individual inspection of the interface.

2.3. Thinking Aloud Study (TA)

Since TA was first introduced in systems development, numerous variations of the method have been employed, and today there is no definitive definition of the aim and usage of the method and no single accepted procedure to follow. TA is used in various situations—with various goals—both early and late in the development cycle (see Dumas & Redish, 1993; Nielsen, 1994b). TA can, for example, be performed by usability specialists in a usability laboratory with video cameras and one-way

---

[2]The heuristics have been slightly reworded to transform them from headings ("error prevention") to instructions ("prevent errors").

mirrors, or it can be performed in the field and analyzed on the fly by systems developers. The common core of TA is that it involves a small number of users who think out loud while solving tasks with the system that is being tested, and an evaluator who detects usability problems by observing the users and listening in on their thoughts. It is generally held that at least four to five users are necessary to detect the majority of the problems in a system, but the necessary number of users varies considerably with the aim of the test and the complexity and quality of the system (see J. R. Lewis, 1994).

The general procedure for TA consists of a preparation phase followed by a number of test sessions, normally one for each user. In the preparation phase, the people conducting the test familiarize themselves with the work environment where the system is going to be used, define appropriate tasks, and recruit users. The test sessions are administered by a facilitator, who may at the same time be the person evaluating when the users experience problems. Each session consists of an introduction to familiarize the user with the test situation, the actual test, and a debriefing of the user. In the introduction, the facilitator should teach the user to think out loud because this is an unnatural thing to do for users, and experience indicates that without teaching—and some encouragement during the session—only few users are capable of giving valuable verbal reports about their work. The actual test is initiated by reading the first task out loud and handing it over to the user who solves it while thinking out loud. After finishing the first task, the second is presented in a similar manner, and so forth. When the user has finished all tasks, or when time runs out, the user is debriefed to provide any additional insights into the system and to relax after the test session. After all test sessions have been run, the evaluator produces a complete, nonduplicate list of the detected problems.

3. MEASURING THE EVALUATOR EFFECT

Previous studies (e.g., Hertzum & Jacobsen, 1999; Jacobsen, Hertzum, & John, 1998; Nielsen, 1992) have used the average detection rate of a single evaluator as their basic measure of the evaluator effect. This measure relates the evaluators' individual performances to their collective performance by dividing the average number of problems detected by a single evaluator by the number of problems detected collectively by all the evaluators. These calculations are based on unique problems—that is, after duplicate problems within and between evaluators have been eliminated (see Equation 1).

$$\textit{Detection rate} = \text{Average of } \frac{|P_i|}{|P_{all}|} \text{ over all } n \text{ evaluators} \tag{1}$$

In the equation, P_i is the set of problems detected by evaluator i (i.e., the problem list of evaluator i) and P_{all} is the set of problems detected collectively by all n evaluators. The detection rate is easy to calculate, and it is available for the 11 studies reviewed in this article. However, the detection rate suffers from two drawbacks. First, the lower bound for the detection rate varies with the number of evaluators. Using only one evaluator, the detection rate will always be 100%, using two evaluators it will be at least 50% (when there is no overlap between the two evaluators' problem lists), and using n evaluators it will be at least $100/n$ percent. When the number of evaluators is small, it is important to interpret the detection rate as a value between the lower bound and 100%, not between 0% and 100%, otherwise the detection rate will appear higher than it actually is. Second, and related, the detection rate rests on the assumption that the number of problems found collectively by the evaluators is identical to the total number of problems in the interface. A small group of evaluators is, however, likely to miss some problems and then the detection rate becomes overly high. Adding more evaluators will normally lead to the detection of some hitherto missed problems, and this improvement of the evaluators' collective performance is reflected in the detection rate as a decrease in the average performance of individual evaluators.

To avoid the problems caused by relating the performance of single evaluators to the collective performance of all evaluators, the *any-two agreement* measures to what extent pairs of evaluators agree on what problems the system contains. The any-two agreement is the number of problems two evaluators have in common divided by the number of problems they collectively detect, averaged over all possible pairs of two evaluators (see Equation 2).

$$Any\text{-}two\ agreement = \text{Average of } \frac{|P_i \cdot P_j|}{|P_i \cdot\cdot P_j|} \text{ over all } 1/2n(n-1) \text{ pairs of evaluators} \qquad (2)$$

In the equation, P_i and P_j are the sets of problems detected by evaluator i and evaluator j, and n is the number of evaluators. The any-two agreement ranges from 0% if no two evaluators have any problem in common to 100% if all evaluators have arrived at the same set of problems. It should be noted that the any-two agreement measures agreement only. A high any-two agreement is no guarantee that all, or even most, problems in the interface have been detected. The any-two agreement is our preferred measure of the evaluator effect,[3] but we have the data necessary to calculate it for only a subset of the studies we review in the next section.

4. STUDIES OF THE EVALUATOR EFFECT IN CW, HE, AND TA

Reviewing the UEM literature, we found six studies that explicitly aimed at investigating the evaluator effect (Hertzum & Jacobsen, 1999; Jacobsen et al., 1998; Jacobsen & John, 2000; Nielsen, 1992, 1994a; Nielsen & Molich, 1990). Two additional

---

[3]The Kappa statistic is often used for measuring interrater agreement. However, Kappa presupposes—like the detection rate—that the total number of problems in the interface is known or can be reliably estimated. Because this is not the case for most of the studies reviewed in this article (the number of evaluators is too small), we prefer to use the any-two agreement.

studies (Molich et al., 1998, 1999) touched on the evaluator effect in a broader sense. In addition to that, one study (C. Lewis et al., 1990), which did not particularly aim at investigating the evaluator effect, contained data that enabled us to investigate it. Finally, the authors of two UEM studies generously provided us with additional data that enabled us to investigate the evaluator effect in their studies (Connell & Hammond, 1999; Dutt, Johnson, & Johnson, 1994).

We would have liked our review to include studies in which CW was performed by groups of cooperating evaluators because this has been suggested as an improvement of the method (Wharton et al., 1992, 1994). We would also have welcomed studies in which HE was performed by evaluators who aggregated the results of their individual inspections to a group output because this is how HE is described. We have, however, only been able to find studies where CW and HE were performed by evaluators working individually. It is currently unknown what effect collaboration among evaluators has on the reliability of usability evaluations. For software inspections, it was found that having the inspectors meet to aggregate the results of their individual inspections leads to the detection of few new defects but does lead to the loss of a number of defects that were originally detected by individual inspectors—meeting losses outweigh meeting gains (Miller, Wood, & Roper, 1998).

4.1. Overview

Table 1 shows that substantial evaluator effects were found for all three UEMs across a range of experimental settings. Specifically, the evaluator effect is neither restricted to novice evaluators nor to evaluators knowledgeable of usability in general. The evaluator effect was also found for evaluators with experience in the specific UEM they have been using (Jacobsen et al., 1998; Lewis et al., 1990; Molich et al., 1998, 1999). Furthermore, the evaluator effect is not affected much by restricting the set of problems to only the severe problems. The first three columns in Table 1 contain information on which study is reviewed, the UEM investigated, and the type of system evaluated. The fourth column indicates whether the system was tested against set tasks, which may have had an impact on the evaluator effect because they tend to make the test sessions more similar. The fifth column gives the total number of unique problems detected by the evaluators collectively. Duplicated problems were eliminated from this number; that is, when an evaluator detected the same problem twice or when two evaluators detected the same problem, this problem was only counted once. The sixth column gives the number and profile of the evaluators. The seventh column gives the detection rate (see Equation 1). For example, in the study by C. Lewis et al. (1990), an evaluator, on average, detected 65% of the total number of problems detected collectively by the four evaluators. In some but not all studies, a set of severe problems was extracted from the total set of problems. The eighth column gives the detection rate for severe problems only, as it is interesting to know whether the evaluator effect was smaller for severe problems than for all problems. For example, in the study by Hertzum and Jacobsen (1999), an evaluator detected, on average, 21% of the severe problems, which was only slightly more than when calcu-

Table 1: Summary of Results in the 11 Reviewed Studies

Reference	UEM	Evaluated System	Task Scenarios	Total Problems Detected	Evaluators	Detection Rate, All Problems (%)	Detection Rate, Severe Problems (%)	Any-Two Agreement
Lewis, Polson, Wharton, & Rieman (1990)	CW	E-mail	Yes	20	4 (3 of the developers of CW and a CE+ novice)	65	—	—
Dutt, Johnson, & Johnson (1994)[a]	CW	Personnel recruitment	Yes	32	3 (2 CS graduate students and an HCI researcher)	73	—	65
Hertzum & Jacobsen (1999)	CW	Web-based library	Yes	33	11 CS graduate students	18	21	17
Jacobsen & John (2000)	CW	Multimedia authoring	No	46	2 CS graduate students	53	—	6
Nielsen & Molich (1990)[b]	HE	Savings	No	48	34 CS students	26	32	26
		Transport	No	34	34 CS students	20	32	—
		Teledata	No	52	37 CS students	51	49	45
		Mantel	No	30	77 computer professionals	38	44	
Nielsen (1992)[c]	HE	Banking	No	16	31 novices (CS students)	22	29	33
					19 usability specialists	41	46	
					14 double specialists	60	61	—
Nielsen (1994a)[b]	HE	Integrating	Yes	40	11 usability specialists	29	46	—
Connell & Hammond (1999)[d]	HE	Hypermedia browser	No	33	8 undergraduates[e]	18	19	9
				84	5 HCI researchers[e]	24	22	5
		Interactive teaching		57	8 psychology undergraduates[e]	20	16	8
Jacobsen, Hertzum, & John (1998)	TA	Multimedia authoring	Yes	93	4 HCI researchers with TA experience	52	72	42
Molich et al. (1998)	TA	Electronic calendar	No	141	3 commercial usability laboratories[f]	37	—	6
Molich et al. (1999)[g]	TA	Web-based e-mail	No	186	6 usability laboratories[h]	22	43	7

Note. The detection rates for all problems and for severe problems only should not be compared across studies without having a closer look at the methodology used in each of the studies. A dash (—) indicates that the figure could not be calculated from the available data. UEM = usability evaluation method; CW = cognitive walkthrough; CE+ = cognitive theory of exploratory learning; CS = computer science; HCI = human–computer interaction; HE = heuristic evaluation; TA = thinking-aloud study.

[a]Hilary Johnson generously gave us access to the data set from Dutt et al. (1994). [b]The detection rates for severe problems are reported in Nielsen (1992). [c]The any-two agreement is calculated on the basis of data reported in Nielsen (1994a). [d]Iain Connell generously gave us access to additional data from Connell & Hammond (1999). [e]More evaluators participated in the study by Connell & Hammond (1999). We have extracted those using the 10 HE heuristics. [f]Four teams participated in the study by Molich et al. (1998), but only three of them used TA. [g]Rolf Molich has generously made the data from the study available at http://www.dialogdesign.dk/cue.html [h]Nine teams participated in the study by Molich et al. (1999), but only six of them used TA.

lating the detection rate for the full set of known problems. The last column gives the any-two agreement (see Equation 2) for the studies in which we were able to calculate it. For example, in the study by Connell and Hammond (1999), the average agreement between any two evaluators was 5% to 9%.

As described in Section 3, a study with few evaluators is likely to yield an overly high detection rate because some problems remain unnoticed by all the evaluators. This leads us to assume that adding more evaluators to the studies that originally involved only a few evaluators will cause a drop in the achieved detection rates (see J. R. Lewis, this issue, for a formula for estimating the drop in detection rate based on data from the first few evaluators). Thus, an overall estimate of the detection rate should probably lean toward the studies with the larger number of evaluators, and these studies generally report the lower detection rates.

4.2. Closer Look at the Reviewed Studies

To provide the detail necessary to assess the credibility of the reviewed studies, in this section we point out the special characteristics of the individual studies and describe three studies, one for each UEM, in more detail.

CW

C. Lewis et al. (1990) had four evaluators individually perform a CW and found that the evaluators were fairly consistent and that they collectively detected almost 50% of the problems revealed by an empirical evaluation with 15 users. The generalizability of these results is difficult to assess, however, because three of the evaluators were creators of CW and had discussed the general trends of the empirical evaluation prior to their walkthroughs (C. Lewis et al., 1990, pp. 238–239). Dutt et al. (1994) had three evaluators individually perform a CW, and they were unusually consistent. Dutt et al. noted that "the number of problems found is relatively low given the quality of the interface" (p. 119). This could indicate that the evaluators did not find all the problems in the interface or that they applied a rather high threshold for the amount of difficulty or inconvenience inflicted on the user before they reported a problem. The studies of C. Lewis et al. (1990) and Dutt et al. were based on the first version of CW, which used a single-page form with nine general questions and several subquestions to evaluate each action. Jacobsen and John (2000) studied two novice evaluators as they learned and used the simpler, current version of CW. Based on a detailed system-specification document, the evaluators individually selected the tasks to be evaluated and spent 22 to 25 hr evaluating the system. The agreement between the evaluators was disappointingly low with only 6% of the problems being detected by both evaluators.

Hertzum and Jacobsen (1999) had 11 first-time users of CW evaluate a Web-based library system against three set tasks. Thus, the study bypassed task selection to focus on the construction and walkthrough of the action sequences. The evaluators, who were graduate students in computer science, received 2 hr of instruction in the CW technique. This instruction consisted of a presentation of the practitioner's guide to CW (Wharton et al., 1994), a discussion of this guide, and an exercise in which the evaluators got some hands-on experience and instant feedback. As a rough estimate, each evaluator spent 2 to 3 hr individually completing his CW, which had to be documented in a problem list describing each detected problem and the CW question that uncovered it. Based on the 11 problem lists, the two authors of Hertzum and Jacobsen (1999) independently constructed a master list of unique problem tokens. The authors agreed on 80% of the problem tokens and resolved the rest through discussion. The evaluators differed substantially with respect to which and how many problems they detected. The largest number of problems detected by a single evaluator was 13, whereas the lowest was 2. As much as 58% of the 33 problems detected collectively by the evaluators were only detected once, and no single problem was detected by all evaluators.

HE

Nielsen and Molich (1990) reported on the HE of four simple walk-up-and-use systems. Two of the systems (Savings and Transport) were running versions of voice-response systems; the two other systems (Teledata and Mantel) were evaluated on the basis of screen dumps. In all four evaluations it was found that aggregating the findings of several evaluators had a drastic effect in the interval from 1 to about 5 evaluators. After that, the effect of using an extra evaluator decreased rapidly and seemed to reach the point of diminishing returns at aggregates of about 10 evaluators. In a second study, Nielsen (1992) compared three groups of evaluators who performed an HE of a simple system giving people access to their bank accounts. The three groups of evaluators were novices, regular usability specialists, and specialists in both voice-response systems and usability (the double specialists). The performance of the evaluators, who made their evaluation from a printed dialogue that had been recorded from the system, increased with their expertise. However, even the double specialists displayed a notable evaluator effect, and they differed just as much in their detection of severe problems as they did for problems in general. In a third study, Nielsen (1994a) reported on an HE of a prototype of a rather complex telephone company application intended for a specialized user population. This study provides evidence that HEs of more complex systems are also subject to a substantial evaluator effect.

Connell and Hammond (1999) conducted two experiments to investigate the effect of using different sets of usability principles in usability evaluations. We focused on the evaluators using the 10 HE heuristics as their usability principles. In the first experiment, a group of novice evaluators and a group of evaluators with human–computer interaction (HCI) knowledge applied HE to a hypermedia browser. Problems were collected by observing the evaluators who were asked to think out loud. In the second experiment, a group of novice evaluators applied HE to an interactive teaching system. Here, the normal HE procedure was followed in that the evaluators reported their findings themselves. The detection rates in both experiments were among the lowest obtained in the studies of HE. Connell and Hammond argued that they were more cautious not to group distinct problems into the same unique problem token and therefore got lower detection rates. The duplicate-elimination process, in which the problem lists of individual evaluators are merged into one master list of unique problem tokens, can result in misleadingly high detection rates if the problems are grouped into too few, overly broad problems. Another explanation could be that the number of problems in the interface increases with the complexity of the system, and this reduces the likelihood that two evaluators will detect the same set of problems. If that is the case, studies of the evaluator effect should be performed on realistically complex systems, such as those used by Connell and Hammond.

TA

There has not been much focus on the evaluator effect in TA. In two independent studies, Molich et al. (1998, 1999) investigated to what extent usability laboratories around the world detect the same problems in a system based on a TA. There are, inarguably, more differences among laboratories than among the evaluators performing the evaluations (e.g., differences in test procedure and different individuals participating as users). We have included these two studies in the review but emphasize that they differ from the other reviewed studies—lower agreement must be expected because more constituents of the evaluation were allowed to vary. In the first study (Molich et al., 1998), three[4] commercial usability laboratories evaluated the same running system with TA based on a two-page description of the primary user group and the aim of the test. As many as 129 of the 141 reported problems were only detected once. In the second study (Molich et al., 1999), six[5] usability laboratories evaluated a commercial Web-based system with TA, this time based on a more precise description of the users and the aim of the test. Again, the laboratories disagreed substantially in that 147 of the 186 reported problems were only detected once.

Only Jacobsen et al. (1998) aimed specifically at revealing the evaluator effect in TA. In this study, four HCI researchers—two with extensive TA experience and two with some TA experience—independently analyzed the same set of videotapes of four usability test sessions. Each session involved a user thinking out loud while solving set tasks in a multimedia authoring system. The evaluators, who also had access to the system and its specification document, were asked to report all problems appearing in the four videotaped test sessions. The evaluators were not restricted in the time they spent analyzing the videotapes, but to minimize individual differences in their conceptions of what constitutes a usability problem, nine set criteria were used. Hence, the evaluators were requested to de-

---

[4]Four teams participated in the study by Molich et al. (1998) but only three of them used TA.
[5]Nine teams participated in the study by Molich et al. (1999) but only six of them used TA.

tect problems according to the nine criteria, and they were asked to report time-stamped evidence and a free-form description for each problem. Based on the evaluators' problem lists, two of the authors of Jacobsen et al. (1998) independently constructed a master list of unique problem tokens. They agreed on 86% of the problem tokens, and by discussing their disagreements and the problems they did not share, a consensus was reached. As much as 46% of the problems were only detected by a single evaluator and another 20% by only two evaluators. Compared to the two less-experienced evaluators, the two evaluators with extensive TA experience spent more time analyzing the videotapes and found more problems. The average detection rate for the two experienced evaluators was 59% but they agreed on only 40% of the problems they collectively detected. The average detection rate for the two less-experienced evaluators was 45% and they agreed on 39% of the problems they collectively detected. Even though the study was set up to minimize the evaluator effect by being more restrictive than most practical TA studies, a substantial evaluator effect remained.

4.3. Severity Judgments

Seven studies, corresponding to 14 experiments, included an assessment of problem severity. In the study by Connell and Hammond (1999), problem severity was assessed on a 7-point rating scale ranging from 1 (*trivial, might be ignored*) to 7 (serious, must be addressed), and severe problems were defined as those receiving one of the three highest rates. The other studies that assessed problem severity did so by dividing the problems into two categories: severe and nonsevere. This was done by having the evaluators point out the problems that ought to be fixed before release of the system (Hertzum & Jacobsen, 1999), by having the evaluators point out the 10 most severe problems (Jacobsen et al., 1998), by stipulating a set of core problems (Molich et al., 1999), or based on expected impact on the user (Nielsen & Molich, 1990; Nielsen, 1992, 1994a). Four experiments displayed an appreciably higher detection rate for severe problems; the other 10 experiments displayed largely no difference between the detection rate for all problems and the detection rate for severe problems only (see Table 1). Thus, the evaluator effect is not merely a disagreement about cosmetic, low-severity problems, which are more or less a matter of taste. For all three UEMs, a single evaluator is unlikely to detect the majority of the severe problems that are detected collectively by a group of evaluators.

Another way of looking at the evaluator effect is to investigate to what extent evaluators agree on what constitutes a severe problem. In a number of the reviewed studies, several of the evaluators were also asked to judge the severity of the problems on the complete list of unique problems. The evaluators' assessments of problem severity are suitable for comparison because they are made independently and based on the same list of problems. The evaluators could, however, be biased toward perceiving the problems they originally detected themselves as more severe than the problems they missed. Lesaigle and Biers (2000) reported a statistically significant bias for 4 of the 13 evaluators who assessed problem severity in their study. Jacobsen et al. (1998) and Nielsen (1994a) also investigated this potential bias and found that it was negligible.

The evaluators in Jacobsen et al. (1998) received the complete list of unique problems with a short description of each unique problem and additional information about, among other things, the number of users experiencing it and the number of evaluators detecting it. Each evaluator was presented with a scenario in which a project manager had constrained the evaluators to point out the 10 most severe problems, as a tight deadline left room for fixing only those few problems in the next release. After they had created their top-10 lists, the evaluators were also asked to write down their strategies for creating their lists. The strategies varied greatly among the evaluators and were based on multiple aspects such as the evaluators' favor for certain user groups, the number of evaluators and users encountering a problem, the violated problem criteria, expectations about real-world usage of the system, and so forth. All these aspects may catch important dimensions of problem severity but they also led the evaluators to select markedly different sets of problems for their top-10 lists.

Table 2, which covers one study for each of the three UEMs, shows the extent to which evaluators who assess problem severity agree on the set of severe problems. The table gives the any-two agreement among the evaluators with respect to which problems they considered severe and the average correlation between the severity ratings provided by any two evaluators. Nielsen (1994a) stated that "the reliability of the severity ratings from single evaluators is so low that it would be advisable not to base any major investment of development time and effort on such single ratings" (pp. 49–50). In Hertzum and Jacobsen (1999), 35% of the total set of severe problems were only rated severe once. In Jacobsen et al. (1998), 56% of the problems on the top-10 lists were only rated severe once. Not a single problem was unanimously judged as severe in these two studies. In sum, Table 2 shows that the CW, HE, or TA performed by the evaluators did not give rise to a common agreement as to what constituted the central usability issues in the interfaces.

Table 2: Three Studies in Which a Group of Evaluators Judged Problem Severity

Reference	UEM	Evaluated System	Evaluators Who Assessed Severity	No. of Severe Problems	Any-Two Agreement on Severity Ratings	Average Spearman Correlation (SD)
Hertzum & Jacobsen (1999)	CW	Web-based library	6 CS graduate students	20	28%	.31 (.18)
Nielsen (1994a)	HE	Integrating	11 usability specialists	—	—	.24
Jacobsen, Hertzum, & John (1998)	TA	Multimedia authoring	4 HCI researchers with TA experience	25	20%	.23 (.16)

Note. A dash (—) indicates that the figure could not be calculated from the available data. UEM = usability evaluation method; CW = cognitive walkthrough; CS = computer science; HE = heuristic evaluation; TA = thinking-aloud study; HCI = human–computer interaction.

5. DISCUSSION

The evaluator effect has been documented for different UEMs, for both simple and complex systems, for both paper prototypes and running systems, for both novice and experienced evaluators, for both cosmetic and severe problems, and for both problem detection and severity judgment. The question is not whether the evaluator effect exists but why it exists and how it can be handled. We believe the principal reason for the evaluator effect is that usability evaluation involves interpretation. Although some usability problems are virtually self-evident, most problems require the evaluator to exercise judgment in analyzing the interaction among the users, their tasks, and the system. It should be noted that evaluator effects are not specific to usability evaluation. Interobserver variability also exists for more mature cognitive activities such as document indexing (e.g., Funk, Reid, & McCoogan, 1983; Sievert & Andrews, 1991; Zunde & Dexter, 1969) and medical diagnosing (e.g., Corona et al., 1996; Cramer, 1997; Sørensen, Hirsch, Gazdar, & Olsen, 1993). In general, individual differences—often categorized into groups such as cognitive abilities, expertise, motivation, personality, and skill acquisition—preclude that cognitive activities such as detecting and assessing usability problems are completely consistent across evaluators.

In analyzing how interpretation enters into usability evaluations and gives rise to differences across evaluators, we focused on where the UEMs fell short of providing evaluators with the guidance necessary for performing reliable evaluations. Three such shortcomings of the UEMs are (a) vague goal analyses leading to the selection of different task scenarios, (b) vague evaluation procedures leading to anchoring, and (c) vague problem criteria leading to anything being accepted as a problem. In addition to discussing what causes the evaluator effect, we also make suggestions regarding how it can be dealt with.

5.1. Vague Goal Analyses

At least for complex systems, it is not practically possible to include all aspects of a system in one evaluation. Consequently, it is important to analyze what the evaluation is to achieve and focus it accordingly. Vague goal analysis prior to usability evaluation leaves many decisions about which aspects of the system to include in the evaluation to the evaluator's discretion. Although evaluators may agree in general on the focus of an evaluation, small differences in their selection of which specific functions to evaluate for different system features may lead to considerable variability in the evaluators' final choice of evaluation tasks. Although evaluating different aspects of the system might not be thought of as an evaluator effect per se, it certainly impacts the results of a usability evaluation.

The outcome of the goal analysis can simply be a mental clarification, but the goal analysis can also result in a set of task scenarios the system is to be evaluated against. HE relies on a merely mental clarification of the goal of the evaluation, CW makes use of task scenarios but includes no guidance on task selection, and the various versions of TA normally include task scenarios devised on the basis of interaction with target users. In 5 of the experiments reviewed in the previous section, the evaluators received identical task scenarios for their evaluation; in the other 13 experiments, task scenarios were either not used at all or it was left to the individual evaluators to select them. In Jacobsen and John (2000), the two CW evaluators were to set up task scenarios by themselves, and 43% of the problems that were detected by one evaluator and missed by the other stemmed from tasks selected and evaluated by only one of the evaluators. The reviewed studies provide ample evidence that the evaluator effect is not eliminated by giving the evaluators identical task scenarios, but it must be suspected that vague goal analyses introduce additional variability.

Task selection, or the broader activity of goal analysis, seems a somewhat neglected aspect of the three reviewed UEMs. We suggest that to reduce the evaluator effect and in general improve the quality of their evaluations, evaluators should verify the coverage of their task scenarios in a systematic way. Such an analysis of task coverage is intended to ensure that all relevant system facilities are considered for inclusion in a task scenario and hence provides a basis for selecting the optimal subset of facilities for actual inclusion in the evaluation. The most important facilities to test will generally be the high-risk ones and those with a known or expected high frequency of use.

5.2. Vague Evaluation Procedures

Whereas TA and in particular CW provide the evaluator with a procedure describing the phases of the evaluation and how to complete them, HE does not offer much in terms of a procedure for driving the evaluation. The heuristics used in HE "seem to describe common properties of usable interfaces" (Nielsen, 1994a, p. 28), but HE does not provide a systematic procedure for ensuring that all interface elements are evaluated against all heuristics. Thus, although the heuristics take one step toward pointing to the problems in the interface, they still leave a considerable gap for the evaluator to close. We conjecture that the heuristics principally serve as a source of inspiration in that they support the evaluator in looking over the interface several times while focusing on different aspects and relations of it. The quality of this stream of new aspects and relations is that it leads the evaluator to consider still new questions about the usability of the interface. Although novice evaluators may use the heuristics in this concurrent and inspirational way, experienced evaluators may chiefly get their inspiration from the experiences they have accumulated during past evaluations. Thus, HE leaves room for using the heuristics in different ways and to different extents. This is a deliberate feature of HE, which is intended as an easily applicable informal method, but it also leads to an evaluator effect.

Although a substantial evaluator effect may not be surprising for a UEM as informal as HE, it is certainly notable that the strict procedure of CW does not lead to consistently better agreement among the evaluators. In a CW, the evaluator specifies a user or a group of users (e.g., "users with Mac® experience"), which is the basis for answering the four questions for each action in the action sequences. It is a critical assumption of CW that posing the four questions helps evaluators reach reliable answers. It is, however, not evident to what extent this is the case. To answer the questions accurately, the evaluator needs to know quite a lot about how the specified users will react to different user interface properties and facilities— knowledge that is not typically made explicit in the user description (Hertzum & Jacobsen, 1999; Jacobsen & John, 2000). In case of insufficient knowledge of how the users will react to the interface, the walk-through becomes inaccurate due to a phenomenon known as *anchoring*; that is, despite the evaluator's efforts, the walk-through ends up evaluating the system against a user who is much too similar to the evaluator to be representative of the actual users. Each of the four questions in CW drives evaluators to think of the user's behavior in a certain situation, but when the general user description becomes too fuzzy, the evaluators unintentionally substitute it with their own experience with the system. The anchoring hypothesis was investigated by Jacobsen and John (2000), who kept track of the evaluators' learning and evaluation process through diaries written by the evaluators themselves. For both evaluators, Jacobsen and John found examples of usability problems the evaluators reported in and credited to their CWs, although the evaluators noticed these problems during their preparation phase up to 15 hr before encountering them as part of their walkthrough process. This illustrates that usability problems experienced personally by evaluators are likely to enter into their evaluations by showing up later as reported usability problems.

The anchoring hypothesis can readily be extended to HE, which is also an inspection method, but one could hope that it would not extend to TA in which the evaluator observes, rather than imagines, target users interacting with the system. However, differences in TA evaluators' general views on usability, their personal experiences with the system under evaluation, their opinions about it, and so forth, lead them to make some observations and remain blind toward others. We can only hypothesize that this is due to anchoring, but the magnitude of the resulting evaluator effect testifies to the considerable amount of interpretation involved in evaluating TA sessions, in spite of the rather concrete procedure.

The effect of adding more evaluators to a TA study resembles the effect of adding more users; both additions increase the overall number of problems found and by comparable amounts. In fact, the study by Jacobsen et al. (1998) suggested that the geometric mean of the number of users and evaluators is a rule-of-thumb estimate of the total number of problems identified in a TA study (see Equation 3). This means that to maximize the number of problems found and, simultaneously, minimize the number of users and evaluators, the number of users and evaluators should be the same: Three evaluators individually observing three users are more productive in identifying usability problems than is one evaluator observing five users. It should be kept in mind that Equation 3 is derived from a study with only four users and four evaluators. This may not be enough to make reliable predictions for large numbers of users and evaluators.

$$\textit{Number of problems found} \gg C\sqrt{\textit{number of evaluators} \cdot \textit{number of users}} \quad (3)$$

If we take as our premise that each evaluator examines the interface once per user, then setting the number of evaluators equal to the number of users maximizes the number of examinations of the interface. Hence, the crucial factor to consider in deciding on how many users and evaluators to involve in a TA study is the number of examinations of the interface. As the time invested and the hourly price for each evaluator is normally higher than for each user, it will probably not be cost effective to have an equal number of users and evaluators but it could be considered to trade a couple of users for an extra evaluator. A positive side effect of this suggestion is that, in comparing their results, the evaluators will have an opportunity to discuss and learn from the nature and size of their disagreements, thus increasing their awareness of the evaluator effect.

Previous studies of how many users to include in TA studies have found that the number of problems detected can be modeled by the formula $N[1 - (1 - p)^u]$, where N is the total number of problems in the interface, p is the probability of finding the average problem when running a single, average user, and u is the number of users participating in the evaluation (J. R. Lewis, 1994; Nielsen & Landauer, 1993; Virzi, 1992). Both Equation 3 and the $N[1 - (1 - p)^u]$ formula predict diminishing returns for increasing numbers of users (and evaluators); that is, adding another user or evaluator will yield fewer and fewer hitherto unnoticed problems as the number of users and evaluators increases. However, Equation 3 rejects the idea of a total number of problems in the interface—rather the number of problems will keep increasing for each new user and evaluator. It should however be reemphasized that Equation 3 may not make reliable predictions for large numbers of users and evaluators.

TA studies with a constant number of evaluators yield diminishing returns for increasing numbers of users, but Equation 3 indicates that whereas a TA study with one evaluator may close in on one value for the total number of problems in the interface, studies with more evaluators will close in on higher values. Thus, if formulas such as $N[1 - (1 - p)^u]$ are used to estimate the total number of problems in an interface based on a TA study performed by a single evaluator, we must expect that the number of problems is underestimated. Figure 1, from Jacobsen et al. (1998), depicts the number of problems detected as a function of the number of both evaluators and users. Each curve corresponds to a fixed number of evaluators. Looking at the evaluators' performance after they had analyzed all four users, the average increase in problems found was 42% going from one to two evaluators, 20% going from two to three evaluators, and 13% going from three to four evaluators.

5.3. Vague Problem Criteria

Heuristics such as "ensure match between system and the real world" do not tell how big a mismatch is allowed to be before it becomes a problem. Similarly, CW provides no guidance on how quickly and effortlessly the user should notice that the correct action is available before this action must be said to be insufficiently noticeable. For TA, it is also uncommon that the evaluators have explicit criteria defining when a difficulty or inconvenience experienced by the user constitutes a usability problem. However, in one of the reviewed studies of TA (Jacobsen et al., 1998), the evaluators were provided with nine predefined criteria defining when an observation should be recorded as a usability problem. Thus, differences in the evaluators'

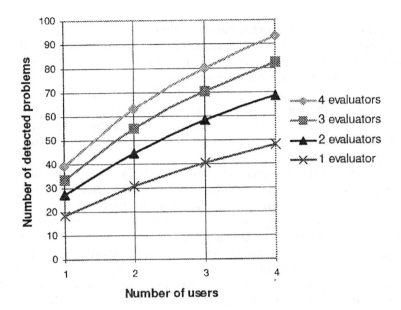

FIGURE 1 The number of problems detected by different numbers of users and evaluators in the think-aloud study by Jacobsen, Hertzum, and John (1998). One evaluator analyzing four users found on average 48 problems, but collectively the four evaluators detected a total of 93 problems. From "Proceedings of the Human Factors and Ergonomics Society, 42nd Annual Meeting", 1998. Copyright 1998 by HFES, all rights reserved. Reprinted with permission.

thresholds regarding when a difficulty or inconvenience becomes a problem are generally not regulated by the UEMs and must be suspected to contribute considerably to the evaluator effect. Evaluators are much more likely to disagree in their choice of threshold—and consequently on whether the difficulty or inconvenience inflicted on the user is sufficiently large to constitute a usability problem—than to hold downright contradictory opinions.

The second version of CW (Polson et al., 1992) made use of extensive criteria for supporting the evaluator in determining whether each of the questions led to the detection of a problem. This level of detail and explicitness was given up in the current version of the method, as several studies described the second version of CW as prohibitively formal and tedious (Rowley & Rhoades, 1992; Wharton et al., 1992). In the current version of CW, however, the evaluator still repeats the same four questions for every action. For moderately complex systems, the number of actions may exceed 100—that is, more than 100 repetitions of the same question, although for different actions. Along with the documentation of the walkthrough, this process indeed becomes tedious. In the study by Hertzum and Jacobsen (1999), several of the evaluators unintentionally skipped an action in the midst of an action sequence. Such slips are likely effects of the tedium of a process that requires the evaluators to meticulously follow a formal procedure. This leads to random differences among the evaluators and illustrates that efforts to reduce the evaluator ef-

fect by providing more formal and complete problem criteria may prove ineffec-
tive. Instead of better performance, the increased formality may introduce slips
and other inconsistencies in the evaluators' behavior.

Without a set of criteria defining what constitutes a usability problem, the
reviewed studies end up accepting any problem report as a usability problem. In
contrast, Nielsen (1993) distinguished between usability problems (i.e., problems
concerning how the system is to be operated) and utility problems (i.e., problems
concerning what the system can do). As researchers, we need precise operational
definitions of core concepts, such as usability problem, to make reliable studies of
UEMs (see Gray & Salzman, 1998). Otherwise, two evaluators may make the same
observations but report them differently because of differences in their understanding of what they are looking for. As practitioners, we are somewhat reluctant
to adopt explicit problem criteria because they may favor agreement among evaluators over detection of all problems of practical importance. That is, a shared understanding of what constitutes a usability problem may not only reduce the evaluator effect but also cause evaluators to systematically miss certain types of
problems. We believe explicit problem criteria can reduce the evaluator effect, especially in TA studies. The development of such criteria is, however, not easy, as they
are both system and task dependent and closely associated to the aim of the evaluation. Moreover, no matter how unambiguously the criteria are defined, applying
them is, in the end, a matter of subjective judgment.

6. CONCLUSION

Based on a review of 11 studies of CW, HE, and TA, we have found that different
evaluators evaluating the same system with one of these methods detect substantially different sets of usability problems in the system. This evaluator effect persists
across differences in system domain, system complexity, prototype fidelity, evaluator experience, problem severity, and with respect to detection of usability problems
as well as assessments of problem severity. In the reviewed studies, the average
agreement between any two evaluators ranged from 5% to 65%, and none of the
UEMs is consistently better than the others. The agreement between two evaluators
is the relation between the number of problems they have in common and the number of problems they have collectively detected. As a measure of the evaluator effect, we prefer the any-two agreement to the more widely reported detection rate
because the detection rate is difficult to interpret correctly and measures coverage
rather than agreement.

We believe that the principal cause for the evaluator effect is that usability evaluation is a cognitive activity, which requires that the evaluators exercise judgment.
Thus, complete agreement among evaluators is unattainable. As we consider usability evaluation pertinent to the development of usable systems, we are, however, concerned about the magnitude of the evaluator effect in currently available
UEMs. A substantial evaluator effect may not be surprising for a UEM as informal
as HE, but it is certainly notable that only marginally better agreement among the
evaluators is achieved by adding the strict procedure of CW and by observing us-

ers who think out loud. Three aspects of the methods are considered as contributors to the evaluator effect: (a) vague goal analysis, (b) vague evaluation procedures, and (c) vague problem criteria. Several of the reviewed studies have dealt with one of the three vaguenesses and can serve to illustrate that as long as the other vaguenesses remain, the evaluator effect is still substantial. A couple of the studies attempted to deal with all three vaguenesses and achieved some of the most consistent results, though better agreement must still be a top priority.

6.1. Open Research Questions

Do UEMs produce valid results?   The evaluator effect makes it apparent that evaluators disagree on what problems an interface contains, but it does not tell whether this is due to real problems that are not reported (misses) or reported problems that are not real (false alarms). In most evaluations of UEMs, the issue of false alarms receives no consideration, as any problem report is accepted as a usability problem. This leaves us virtually without evidence on which of the reported problems that matter to actual users doing real work (see also the discussion in Gray & Salzman, 1998; Olson & Moran, 1998). Specifically, we do not know whether evaluators should be advised to apply a higher threshold before they report a problem—to avoid false alarms—or a lower threshold to avoid misses.

Is the evaluator effect a result of interevaluator variability or intraevaluator variability?   We need to investigate whether the evaluator effect reflects a true disagreement among evaluators or owes to inconsistencies in individual evaluators' performance. None of the reviewed studies have investigated whether evaluators are consistent across evaluations. Hence, the evaluator effect as discussed in this study comprises interevaluator variability as well as intraevaluator variability, and we do not know how much each contributes to the overall evaluator effect. The distinction between these two types of variability may be important because they may have different causes.

6.2. Consequences for Practitioners

Be explicit on goal analysis and task selection.   Even for moderately complex systems, it is prohibitively demanding in time and resources to evaluate all aspects of a system in one test. Thus, before doing any usability evaluation, we should thoroughly analyze the goals of the evaluation and carefully select task scenarios: What should this particular evaluation tell us? What aspects of the system should be covered? What should these parts of the system support the user in doing? Who will use the system, and in what contexts? After the task scenarios have been made, their coverage should be checked, and if necessary, the scenarios should be iteratively improved. This process is intended to both strengthen the preparation phase to increase the impact of the evaluation and to ensure agreement among the involved parties as to what the evaluation is to achieve.

Involve an extra evaluator, at least in critical evaluations.   If it is important to the success of the evaluation to find most of the problems in a system, then we strongly recommend using more than one evaluator. For TA, and possibly other user-involving UEMs, it seems that a reduction in the number of users can somewhat compensate for the cost of extra evaluators without degrading the quality of the evaluation. Further, the multiple evaluators can work in parallel and thus may save calendar time compared to a single evaluator because the single evaluator needs to run more users. Having just two evaluators will both improve the robustness of the evaluation and provide an opportunity for the evaluators to experience for themselves to what extent they disagree and on what types of issues.

Reflect on your evaluation procedures and problem criteria.   The currently available UEMs are not as reliable as we would like them to be. Hence, much is left to personal judgment and work routines established among colleagues. This means that much can be learned from periodically taking a critical look at one's practices to adjust the evaluation procedure, tighten up problem criteria, and so forth. Peer reviewing how colleagues perform usability evaluations seems a valuable source of input for discussions of best practices and a way of gradually establishing a shared notion of usability and usability problems.

Finally, in spite of the evaluator effect, usability evaluations are a prerequisite for working systematically with, ensuring, and improving the usability of computer systems. Although the UEMs reviewed in this article are not perfect, we still believe they are among the best techniques available.

REFERENCES

Connell, I. W., & Hammond, N. V. (1999). Comparing usability evaluation principles with heuristics: Problem instances vs. problem types. In M. Angela Sasse & C. Johnson (Eds.), *Proceedings of the IFIP INTERACT '99 Conference on Human–Computer Interaction* (pp. 621–629). Amsterdam: IOS.

Corona, R., Mele, A., Amini, M., De Rosa, G., Coppola, G., Piccardi, P., Fucci, M., Pasquini, P., & Faraggiana, T. (1996). Interobserver variability on the histopathologic diagnosis of cutaneous melanoma and other pigmented skin lesions. *Journal of Clinical Oncology, 14,* 1218–1223.

Cramer, S. F. (1997). Interobserver variability in dermatopathology. *Archives of Dermatology, 133,* 1033–1036.

Dumas, J. S., & Redish, J. C. (1993). *A practical guide to usability testing.* Norwood, NJ: Ablex.

Dutt, A., Johnson, H., & Johnson, P. (1994). Evaluating evaluation methods. In G. Cockton, S. W. Draper, & G. R. S. Weir (Eds.), *People and computers IX* (pp. 109–121). Cambridge, England: Cambridge University Press.

Funk, M. E., Reid, C. A., & McCoogan, L. S. (1983). Indexing consistency in MEDLINE. *Bulletin of the Medical Library Association, 71*(2), 176–183.

Gray, W. D., & Salzman, M. C. (1998). Damaged merchandise? A review of experiments that compare usability evaluation methods. *Human–Computer Interaction, 13*, 203–261.

Hertzum, M., & Jacobsen, N. E. (1999). The evaluator effect during first-time use of the cognitive walkthrough technique. In H.-J. Bullinger & J. Ziegler (Eds.), *Human–computer interaction: Ergonomics and user interfaces* (Vol. 1, pp. 1063–1067). London: Lawrence Erlbaum Associates, Inc.

Jacobsen, N. E., Hertzum, M., & John, B. E. (1998). The evaluator effect in usability studies: Problem detection and severity judgments. In *Proceedings of the Human Factors and Ergonomics Society 42nd Annual Meeting* (pp. 1336–1340). Santa Monica, CA: Human Factors and Ergonomics Society.

Jacobsen, N. E., & John, B. E. (2000). *Two case studies in using cognitive walkthrough for interface evaluation* (CMU Technical Report No. CMU–CS–00–132). Pittsburgh, PA: Carnegie Mellon University.

Lesaigle, E. M., & Biers, D. W. (2000). Effect of type of information on real time usability evaluation: Implications for remote usability testing. In *Proceedings of the XIVth Triennial Congress of the International Ergonomics Association and 44th Annual Meeting of the Human Factors and Ergonomics Society* (pp. 6-585–6-588). Santa Monica, CA: Human Factors and Ergonomics Society.

Lewis, C. (1982). *Using the "thinking-aloud" method in cognitive interface design* (IBM Research Rep. No. RC 9265 [#40713]). Yorktown Heights, NY: IBM Thomas J. Watson Research Center.

Lewis, C., Polson, P., Wharton, C., & Rieman, J. (1990). Testing a walkthrough methodology for theory-based design of walk-up-and-use interfaces. In *Proceedings of the ACM CHI '90 Conference* (pp. 235–242). New York: ACM.

Lewis, C., & Wharton, C. (1997). Cognitive walkthroughs. In M. Helander, T. K. Landauer, & P. Prabhu (Eds.), *Handbook of human–computer interaction* (Rev. 2nd ed., pp. 717–732). Amsterdam: Elsevier.

Lewis, J. R. (1994). Sample sizes for usability studies: Additional considerations. *Human Factors, 36*, 368–378.

Miller, J., Wood, M., & Roper, M. (1998). Further experiences with scenarios and checklists. *Empirical Software Engineering, 3*(1), 37–64.

Molich, R., Bevan, N., Curson, I., Butler, S., Kindlund, E., Miller, D., & Kirakowski, J. (1998). Comparative evaluation of usability tests. In *Proceedings of the Usability Professionals Association 1998 Conference* (pp. 189–200). Chicago: UPA.

Molich, R., Thomsen, A. D., Karyukina, B., Schmidt, L., Ede, M., van Oel, W., & Arcuri, M. (1999). Comparative evaluation of usability tests. In *Extended Abstracts of ACM CHI '99 Conference* (pp. 83–84). New York: ACM.

Nielsen, J. (1992). Finding usability problems through heuristic evaluation. In *Proceedings of the ACM CHI '92 Conference* (pp. 373–380). New York: ACM.

Nielsen, J. (1993). *Usability engineering.* Boston: Academic.

Nielsen, J. (1994a). Heuristic evaluation. In J. Nielsen & R. L. Mack (Eds.), *Usability inspection methods* (pp. 25–62). New York: Wiley.

Nielsen, J. (Ed.). (1994b). Usability laboratories [Special issue]. *Behaviour & Information Technology, 13*(1 & 2).

Nielsen, J., & Landauer, T. K. (1993). A mathematical model of the finding of usability problems. In *Proceedings of the INTERCHI '93 Conference* (pp. 206–213). New York: ACM.

Nielsen, J., & Molich, R. (1990). Heuristic evaluation of user interfaces. In *Proceedings of the ACM CHI '90 Conference* (pp. 249–256). New York: ACM.

Olson, G. M., & Moran, T. P. (Eds.). (1998). Commentary on "Damaged merchandise?" *Human–Computer Interaction, 13,* 263–323.

Polson, P., & Lewis, C. (1990). Theory-based design for easily learned interfaces. *Human–Computer Interaction, 5,* 191–220.

Polson, P., Lewis, C., Rieman, J., & Wharton, C. (1992). Cognitive walkthroughs: A method for theory-based evaluation of user interfaces. *International Journal of Man–Machine Studies, 36,* 741–773.

Rowley, D. E., & Rhoades, D. G. (1992). The cognitive jogthrough: A fast-paced user interface evaluation procedure. In *Proceedings of the ACM CHI '92 Conference* (pp. 389–395). New York: ACM.

Sievert, M. C., & Andrews, M. J. (1991). Indexing consistency in *Information Science Abstracts. Journal of the American Society for Information Science, 42*(1), 1–6.

Sørensen, J. B., Hirsch, F. R., Gazdar, A., & Olsen, J. E. (1993). Interobserver variability in histopathologic subtyping and grading of pulmonary adenocarcinoma. *Cancer, 71,* 2971–2976.

Virzi, R. A. (1992). Refining the test phase of usability evaluation: How many subjects is enough? *Human Factors, 34,* 457–468.

Wharton, C., Bradford, J., Jeffries, R., & Franzke, M. (1992). Applying cognitive walkthroughs to more complex user interfaces: Experiences, issues, and recommendations. In *Proceedings of the ACM CHI '92 Conference* (pp. 381–388). New York: ACM.

Wharton, C., Rieman, J., Lewis, C., & Polson, P. (1994). The cognitive walkthrough method: A practitioner's guide. In J. Nielsen & R. L. Mack (Eds.), *Usability inspection methods* (pp. 105–140). New York: Wiley.

Zunde, P., & Dexter, M. E. (1969). Indexing consistency and quality. *American Documentation, 20,* 259–267.

INTERNATIONAL JOURNAL OF HUMAN–COMPUTER INTERACTION, *13*(4), 445–479
Copyright © 2001, Lawrence Erlbaum Associates, Inc.

Evaluation of Procedures for Adjusting Problem-Discovery Rates Estimated From Small Samples

James R. Lewis
IBM Corporation

There are 2 excellent reasons to compute usability problem-discovery rates. First, an estimate of the problem-discovery rate is a key component for projecting the required sample size for a usability study. Second, practitioners can use this estimate to calculate the proportion of discovered problems for a given sample size. Unfortunately, small-sample estimates of the problem-discovery rate suffer from a serious overestimation bias. This bias can lead to serious underestimation of required sample sizes and serious overestimation of the proportion of discovered problems. This article contains descriptions and evaluations of a number of methods for adjusting small-sample estimates of the problem-discovery rate to compensate for this bias. A series of Monte Carlo simulations provided evidence that the average of a normalization procedure and Good–Turing (Jelinek, 1997; Manning & Schutze, 1999) discounting produces highly accurate estimates of usability problem-discovery rates from small sample sizes.

1. INTRODUCTION

1.1. Problem-Discovery Rate and Its Applications

Many usability studies (such as scenario-based usability evaluations and heuristic evaluations) have as their primary goal the discovery of usability problems (Lewis, 1994). Practitioners use the discovered problems to develop recommendations for the improvement of the system or product under study (Norman, 1983). The problem-discovery rate (p) for a usability study is, across a sample of participants, the average of the proportion of problems observed for each participant (or the average of the proportion of participants experiencing each observed problem).

It is important to keep in mind that no single usability method can detect all possible usability problems. In this article, reference to the proportion of problems detected means the number of problems detected over the number of detectable prob-

---

Requests for reprints should be sent to James R. Lewis, IBM Corporation, 8051 Congress Avenue, Suite 2227, Boca Raton, FL 33487. E-mail: jimlewis@us.ibm.com

lems. Furthermore, the focus is solely on the frequency of problem occurrence rather than severity or impact. Despite some evidence that the discovery rate for very severe problems is greater than that for less severe problems (Nielsen, 1992; Virzi, 1992), it is possible that this effect depends on the method used to rate severity (Lewis, 1994) or the expertise of the evaluators (Connell & Hammond, 1999).

Discovering usability problems is fundamentally different from discovering diamonds in a diamond mine. The diamonds in a mine may be well hidden, but a group of skilled miners would have excellent agreement about the methods to use to discover them and would not need to spend time discussing whether a particular object in the mine was or was not a diamond. Usability problems, on the other hand, do not exist as a "set of objectively defined, nicely separated problems just waiting for discovery" (M. Hertzum, personal communication, October 26, 2000). Different evaluators can disagree about the usability problems contained in an interface (Hertzum & Jacobsen, this issue). Reference to the number of detectable problems in this article means nothing more than the number of detectable problems given the limitations of a specific usability evaluation setting.

The number of detectable problems can vary as a function of many factors, including but not limited to the number of observers, the expertise of observers, the expertise of participants, and the specific set of scenarios-of-use in problem-discovery observational studies. In heuristic evaluations, the expertise of evaluators and specific set of heuristics can affect the number of detectable problems. For either observational or heuristic methods, the stage of product development and degree of implementation (e.g., paper prototype vs. full implementation) can also affect the number of detectable problems. By whatever means employed, however, once investigators have a set of usability problems in which they can code the presence and absence of problems across participants or observers as a series of 0s and 1s, those sets have some interesting properties.

A hypothetical example.   Suppose a usability study of 10 participants performing a set of tasks with a particular software application (or 10 independent evaluators in a heuristic evaluation) had the outcome illustrated in Table 1.

Because there are 10 problems and 10 participants, the table contains 100 cells. An "x" in a cell indicates that the specified participant experienced the specified problem. With 50 cells filled, the estimated problem-discovery rate (also known as the average likelihood of problem detection) is .50 (50/100). (Note that the averages of the elements in the Proportion column and the Proportion row in the table are both .50.)

Projection of sample size requirements for usability studies.   A number of large-sample usability and heuristic studies (Lewis, 1994; Nielsen & Landauer, 1993; Virzi, 1992) have shown that usability practitioners can use Equation 1 to project problem discovery as a function of the sample size n (number of participants or evaluators) and the problem-discovery rate p.

$$1 - (1 - p)^n \tag{1}$$

Table 1: Hypothetical Results for a Problem-Discovery Usability Study

Participant	Prob 1	Prob 2	Prob 3	Prob 4	Prob 5	Prob 6	Prob 7	Prob 8	Prob 9	Prob 10	Count	Proportion
1	x	x		x		x		x		x	6	0.6
2	x	x		x		x		x			5	0.5
3	x	x		x	x	x					5	0.5
4	x	x		x			x				4	0.4
5	x	x	x	x		x			x		6	0.6
6	x	x	x					x			4	0.4
7	x	x	x		x						4	0.4
8	x	x	x		x		x				5	0.5
9	x		x		x		x		x		5	0.5
10	x		x		x		x		x	x	6	0.6
Count	10	8	6	5	5	4	4	3	3	2	50	
Proportion	1.0	0.8	0.6	0.5	0.5	0.4	0.4	0.3	0.3	0.2		0.50

Note. Prob = problem; x = specified participant experienced specified problem.

447

It is possible to derive Equation 1 from either the binomial probability formula (Lewis, 1982, 1994) or the constant probability path independent Poisson process model (T. K. Landauer, personal communication, December 20, 1999; Nielsen & Landauer, 1993). Regardless of derivational perspective, an accurate estimate of p is essential for the accurate projection of cumulative problem discovery as a function of n, which is in turn essential for the accurate estimation of the sample size required to achieve a given problem-discovery goal (e.g., 90% problem discovery). Figure 1 illustrates projected problem-discovery curves for a variety of values of p.

Estimation of the proportion of discovered problems as a function of sample size. There are times when a usability practitioner does not have full control over the sample size due to various time or other resource constraints. In those cases, the practitioner might want to estimate the proportion of problems detected from the full set of problems available for detection by the method employed by the practitioner. Suppose the practitioner has observed the first 3 participants and has obtained the results presented in the first three rows of Table 1. Because $p = .5$ and $n = 3$, the estimated proportion of problems detected is .875, or $1 - (1 - .5)^3$.

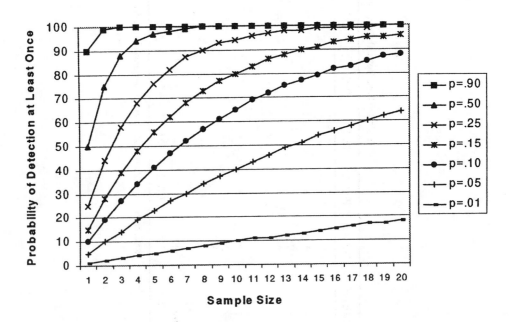

FIGURE 1 Projected problem-discovery curves as a function of n for various values of p.

1.2. Small-Sample Overestimation of the Problem-Discovery Rate

In the preceding example, however, how does the practitioner know that the problem-discovery rate is .5? Observation of the first 3 participants only uncovers Problems 1, 2, 4, 5, 6, 8, and 10, with the participant-to-problem distribution shown in Table 2. Because there are 21 cells (7 observed problems × 3 participants), with 16 of the cells containing an "x," the estimated value of p is .76, about a 50% overestimation relative to the full set of data presented in Table 1. Given this estimate of p, a practitioner would conclude that the study had uncovered 98.6% of the detectable problems.

Hertzum and Jacobsen (this issue) were the first to identify the problem of overestimating p from small-sample usability studies. They pointed out that the smallest possible value of p estimable from a given study is $1/n$, where n is the number of participants in the study. If a study has only 1 participant, the estimate of p will necessarily be 1.0. If a study has 2 participants, the smallest possible estimate of p, regardless of its true value, is .5. Furthermore, the study will produce this minimum value only when every problem is idiosyncratic (no duplication of any problem across multiple participants). Any duplication necessarily increases the estimate of p.

Goals of this research.   In the remainder of this article, I describe research undertaken to accomplish the following goals:

1. Develop an understanding of the extent to which the overestimation of p is a potential problem for usability practitioners.
2. Investigate different procedures for the adjustment of the initial estimate of p to an estimate closer to the true value of p.
3. For the best adjustment procedures, determine how their overestimation or underestimation of p affects projected sample sizes and deviations from problem-discovery goals.

2. METHOD

In the first part of this section, I provide some background on the different adjustment procedures investigated in this series of experiments. In the remaining

Table 2: Hypothetical Results for a Problem-Discovery Usability Study: First 3 Participants

Participant	Prob 1	Prob 2	Prob 4	Prob 5	Prob 6	Prob 8	Prob 10	Count	Proportion
1	x	x	x		x	x	x	6	0.86
2	x	x	x		x	x		5	0.71
3	x	x	x	x	x			5	0.71
Count	3	3	3	1	3	2	1	16	
Proportion	1.00	1.00	1.00	0.33	1.00	0.67	0.33		0.76

Note.   Prob = problem; x = specified participant experienced specified problem.

sections, I describe the dependent measurements used to evaluate the adjustment procedures and the problem-discovery databases used as source material for the investigations.

2.1. Adjustment Methods

I investigated three approaches for adjusting the initial estimate of p: discounting, normalization, and regression.

Discounting.   One way to reduce the magnitude of overestimation of p is to apply a discounting procedure. There are many discounting procedures, all of which attempt to allocate some amount of probability space to unseen events. Discounting procedures receive wide use in the field of statistical natural language processing, especially in the construction of language models (Manning & Schutze, 1999).

The oldest discounting procedure is LaPlace's law of succession (Jelinek, 1997), sometimes referred to as the "add One" method because you add one to the count for each observation. A common criticism of LaPlace's law is that it tends to assign too much of the probability space to unseen events, underestimating true p (Manning & Schutze, 1999).

A widely used procedure that is more accurate than LaPlace's law is Good–Turing estimation (GT; Jelinek, 1997; Manning & Schutze, 1999). There are a number of paths that lead to the derivation of the GT estimator, but the end result is that the total probability mass reserved for unseen events is $E(N_1)/N$, where $E(N_1)$ is the expected number of events that happen exactly once and N is the total number of events. For a given sample, the usual value used for $E(N_1)$ is the actually observed number of events that occurred once. In the context of a problem-discovery usability study, the events are problems. Applying this to the example shown in Table 2, $E(N_1)$ would be the observed number of problems that happened exactly once (2 in the example) and N would be the total number of problems (7 in the example). Thus, $E(N_1)/N$ is 2/7, or .286. To add this to the total probability space and adjust the original estimate of p would result in $.762/(1 + .286)$, or .592—still an overestimate, but much closer to the true value of p.

For problem-discovery studies, there are other ways to systematically discount the estimate of p by increasing the count in the denominator, such as adding the number of problems that occurred once (Add Ones), the total number of problems observed (Add Probs), or the total number of problem occurrences (Add Occs). Using the example in Table 2, this would result in estimates of .696, or $16/(21 + 2)$; .571, or $16/(21 + 7)$; and .432, or $16/(21 + 16)$, respectively.

Suppose one discount method consistently fails to reduce the estimate of p sufficiently, and a different one consistently reduces it to too great an extent. It is then possible to use simple linear interpolation to arrive at a better estimate of p. In the examples used previously, averaging the Add Occs estimation with the Add Probs estimation results in an estimate of .502, or $(.571 + .432)/2$—the closest estimate in this set of examples to the true p of .500.

Normalization.   In their discussion of the problem of overestimation of p, Hertzum and Jacobsen (this issue) pointed out that the smallest possible value of p from a small-sample problem-discovery study is $1/n$. For estimates based on a small sample size, this limit can produce considerable overestimation of p. With larger sample sizes, the effect of this limit on the lowest possible value of p becomes less important. For example, if a study includes 20 participants, then the lower limit for p is $1/20$, or .05.

With the knowledge of this lower limit determined by the sample size, it is possible to normalize a small-sample estimate of p in the following way. Subtract from the original estimate of p the lower limit, $1/n$. Then, to normalize this value to a scale of 0 to 1, multiply it by $(1 - 1/n)$. For the estimate of p generated from the data in Table 2, the first step would result in the subtraction of .333 from .762, or .429. The second step would be the multiplication of .429 by .667, resulting in .286. In this particular case, the result is a considerable underestimation of true p. It is not clear, though, how serious the underestimation would typically be, so submitting this procedure to more systematic evaluation would be reasonable.

Regression.   Another approach for the estimation of true p from a small sample would be to develop one or more regression models (Cliff, 1987; Draper & Smith, 1966; Pedhazur, 1982). The goal of the models would be to predict true p from information available in the output of a small-sample usability problem-discovery study, such as the original estimate of p, a normalized estimate of p, and the sample size.

2.2. Measurements

I wrote a BASIC program that used a Monte Carlo procedure to sample data from published problem-discovery databases with known values of true p (Lewis, 2000e) to produce the following measurements:

1. Mean value of p.
2. Root mean square error ($RMSE$) for estimated p against true p.
3. The central 50% (interquartile) range of the distribution of p.
4. The central 90% range of the distribution of p.

The program could produce this set of statistics for the unadjusted estimate of p and up to five adjusted estimates (based on 1,000 Monte Carlo iterations for each problem-discovery database at each level of sample size). A preliminary experiment (Lewis, 2000e) confirmed that the programmed Monte Carlo procedure sampled randomly and produced results that were virtually identical to complete factorial arrangement of participants for sample sizes of 2, 3, and 4. The impact of combinatorial expansion (the large number of possible combinations of participants) prevented the use of factorial arrangement for the investigation of larger sample sizes.

2.3. Problem-Discovery Databases

The published problem-discovery databases evaluated in this article were:

1. MACERR (Lewis, 1994; Lewis, Henry, & Mack, 1990): This database came from a scenario-driven problem-discovery usability study conducted to develop usability benchmark values for an integrated office system (word processor, mail application, calendar application, and spreadsheet). Fifteen employees of a temporary employee agency, observed by a highly experienced usability practitioner, completed 11 scenarios-of-use with the system. Participants typically worked on the scenarios for about 6 hr, and the study uncovered 145 different usability problems. The problem-discovery rate (p) for this study was .16. Participants did not think aloud in this study.

2. VIRZI90 (Virzi, 1990, 1992): The problems in this database came from a scenario-driven problem-discovery usability study conducted to evaluate a computer-based appointment calendar. The participants were 20 university undergraduates with little or no computer experience. The participants completed 21 scenarios-of-use under a think-aloud protocol, observed by two experimenters. The experimenters identified 40 separate usability problems. The problem-discovery rate (p) for this study was .36.

3. MANTEL (Nielsen & Molich, 1990): These usability problems came from 76 submissions to a contest presented in the Danish edition of *Computerworld* (excluding data from one submission that did not list any problems). The evaluators were primarily computer professionals who evaluated a written specification (not a working program) for a design of a small information system with which users could dial in to find the name and address associated with a telephone number. The specification contained a single screen and a few system messages, which the participants evaluated using a set of heuristics. The evaluators described 30 distinct usability problems. The problem-discovery rate (p) for this study was .38.

4. SAVINGS (Nielsen & Molich, 1990): For this study, 34 computer science students taking a course in user interface design performed heuristic evaluations of an interactive voice response system (working and deployed) designed to give banking customers information such as their account balances and currency exchange rates. The participants uncovered 48 different usability problems with a problem-discovery rate (p) of .26.

(For figures depicting the MANTEL and SAVINGS databases, see Nielsen & Molich, 1990. For the VIRZI90 database, see Virzi, 1990. The MACERR database is available in the Appendix of this article.)

These were the only large-scale problem-discovery databases available to me for analysis. Fortunately, they had considerable variation in total number of participants, total number of usability problems uncovered, basic problem-discovery rate, and method of execution (observational with and without talk-aloud vs. heuristic, different error classification procedures). Given this variation among the databases, any consistent results obtained by evaluating them stands a good chance of generalizing to other problem-discovery databases (Chapanis, 1988).

3. MONTE CARLO SIMULATIONS OF PROBLEM DISCOVERY

This section of the article contains descriptions of Monte Carlo simulations of problem discovery conducted to

1. Evaluate the degree of overestimation produced by small-sample estimates of p (Section 3.1).
2. Investigate a set of discounting procedures to determine which method best adjusted the initial estimates of p (Section 3.2).
3. Develop a set of regression equations for estimating true p from an initial estimate (Section 3.3).
4. Investigate the normalization procedure and the regression equations to determine which procedure in the set best adjusted the initial estimates of p (Section 3.4).
5. Investigate the effectiveness of adjustment using a combination of GT discounting and normalization (Section 3.5).
6. Replicate and extend the previous findings for the best procedures from each class of procedure (discounting, normalization, and regression) using a greater range of sample sizes (Section 3.6).

3.1. How Serious Is the Small-Sample Overestimation of Problem-Discovery Rates?

The primary purpose of the first Monte Carlo experiment was to investigate the extent to which calculating p from small-sample usability studies results in overestimation.

Table 3: Mean Estimates of Discovery Rates as a Function of Sample Size for Published Databases

Source	True p	N = 2	N = 3	N = 4	N = 5	N = 6
MACERR[a]	.16	0.568	0.421	0.346	0.301	0.269
VIRZI90[b]	.36	0.661	0.544	0.484	0.448	0.425
MANTEL[c]	.38	0.724	0.622	0.572	0.536	0.511
SAVINGS[c]	.26	0.629	0.505	0.442	0.406	0.380

[a]Lewis (1994) and Lewis, Henry, and Mack (1990). [b]Virzi (1990, 1992). [c]Nielsen and Molich (1990).

**Table 4: Mean Estimates of Root Mean Square Error as a Function
of Sample Size for Published Databases**

Source	True p	N = 2	N = 3	N = 4	N = 5	N = 6
MACERR[a]	.16	0.406	0.259	0.185	0.140	0.107
VIRZI90[b]	.36	0.306	0.189	0.130	0.094	0.071
MANTEL[c]	.38	0.354	0.254	0.204	0.167	0.143
SAVINGS[c]	.26	0.377	0.253	0.191	0.155	0.129

[a]Lewis (1994) and Lewis, Henry, and Mack (1990). [b]Virzi (1990, 1992). [c]Nielsen and Molich (1990).

Small-sample estimates of p.   Table 3 shows the mean Monte Carlo estimates of p from the published databases for sample sizes from 2 to 6.

RMSE for unadjusted estimates.   Table 4 shows the average *RMSE* from the
Monte Carlo simulation. The *RMSE* is similar to a standard deviation, but rather
than computing the mean squared difference between each data point and the mean
of a sample (the standard deviation), the computation is the mean squared difference between each data point and true p (based on all participants in a given database). The *RMSE* has the desirable characteristic (for a measure of accuracy) of being
sensitive to both the central tendency and variance of a measure. In other words, if
two measurement methods are equally accurate with regard to the deviation of
their mean from a known true value, the measurement with lower variance will
have the lower *RMSE*. A perfect estimate would have an *RMSE* of 0.

Discussion.   The initial estimates of p for all four databases clearly overesti-
mated the true value of p, with the extent of the overestimation declining as the sample size increased, but still overestimating when $n = 6$. The *RMSE* showed a similar
pattern, with the amount of error declining as the sample size increased, but with
nonzero error remaining when $n = 6$. These data indicate that the overestimation
problem is serious and provide baselines for the evaluation of adjustment procedures. (For a more comprehensive analysis of the data, see Lewis, 2000b).

3.2. Evaluation of Discounting Procedures

The purpose of the second Monte Carlo experiment was to evaluate five different
adjustments based on the discounting procedures discussed previously: GT, Add
Ones (ONES), Add Probs (PROBS), Add Occs (OCCS), and the mean of PROBS and
OCCS (PR–OCC).

Estimates of p.   Table 5 shows the mean Monte Carlo estimates of p (unadjusted and adjusted using the discounting procedures) from the published databases for sample sizes from 2 to 6.

Table 5: Adjusted Discovery Rates as a Function of Sample Size and Discounting Method

Source	True p	Adjustment	N = 2	N = 3	N = 4	N = 5	N = 6
MACERR[a]	.16	NONE	0.568	0.421	0.346	0.301	0.269
		ONES	0.397	0.334	0.294	0.266	0.243
		PROBS	0.378	0.315	0.277	0.251	0.230
		OCCS	0.362	0.296	0.257	0.231	0.212
		PR–OCC	0.372	0.304	0.264	0.237	0.217
		GT	0.305	0.237	0.202	0.181	0.164
VIRZI90[b]	.36	NONE	0.661	0.544	0.484	0.448	0.425
		ONES	0.495	0.463	0.437	0.418	0.405
		PROBS	0.441	0.408	0.387	0.374	0.364
		OCCS	0.397	0.352	0.326	0.309	0.298
		PR–OCC	0.427	0.383	0.356	0.338	0.327
		GT	0.397	0.358	0.340	0.331	0.327
MANTEL[c]	.38	NONE	0.724	0.622	0.572	0.536	0.511
		ONES	0.571	0.549	0.528	0.506	0.489
		PROBS	0.483	0.467	0.458	0.447	0.438
		OCCS	0.419	0.383	0.363	0.348	0.338
		PR–OCC	0.466	0.436	0.418	0.402	0.390
		GT	0.473	0.446	0.431	0.415	0.403
SAVINGS[c]	.26	NONE	0.627	0.502	0.442	0.403	0.381
		ONES	0.458	0.421	0.394	0.371	0.359
		PROBS	0.418	0.377	0.354	0.335	0.326
		OCCS	0.385	0.334	0.306	0.287	0.275
		PR–OCC	0.401	0.355	0.330	0.311	0.301
		GT	0.361	0.318	0.298	0.284	0.280

Note. NONE = no adjustment; ONES = discounted adjustment with the Add Ones method; PROBS = discounted adjustment with the Add Problems method; OCCS = discounted adjustment with the Add Occurences method; PR–OCC = mean of PROBS and OCCS estimates; GT = Good–Turing estimation. [a]Lewis (1994) and Lewis, Henry, and Mack (1990). [b]Virzi (1990, 1992). [c]Nielsen and Molich (1990).

Evaluation of accuracy. I conducted an analysis of variance using *RMSE* as the dependent variable and treating databases as subjects in a within-subjects design. The independent variables were sample size (from 2 to 6) and discounting method (NONE, ONES, PROBS, OCCS, PR–OCC, and GT). The analysis indicated significant main effects of sample size, $F(4, 12) = 22.0$, $p = .00002$ and discounting method, $F(5, 15) = 30.6$, $p = .0000003$, and a significant interaction between these effects, $F(20, 60) = 29.0$, $p = .00035$. Figure 2 illustrates the interaction.

Figure 2 shows that as the sample size increased, accuracy generally increased for all estimation procedures (the main effect of sample size). The lines show reasonably clear separation and relatively less accuracy for NONE, ONES, and PROBS—no discounting and the two procedures that provided the least discounting. The *RMSE*s for OCCS, PR–OCC, and GT were almost identical, especially for sample sizes of 4, 5, and 6. The lines suggest potential convergence at some larger sample size.

A set of planned *t* tests showed that all discounting methods improved estimation accuracy of *p* relative to no discounting at every level of sample size (25

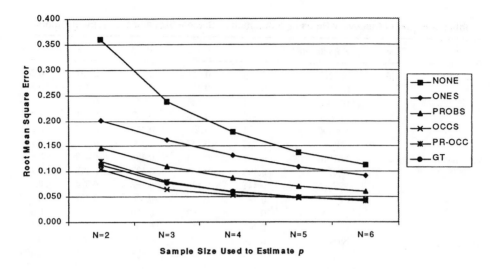

FIGURE 2   Discounting Method × Sample Size interaction for root mean square error.

tests with 3 *df* each, all *ps* < .05). A similar set of planned *t* tests showed that, at all levels of sample size, GT estimation was more accurate than the Add Ones method (5 tests with 3 *df* each, all *p* < .05), but was not significantly more accurate than the other discounting methods (15 tests with 3 *df* each, all *p* > .16). Because no other evaluated discounting procedure was significantly more accurate than GT, all additional analyses involving discounting methods focus on that well-known estimator.

Estimates of required sample size.   GT discounting clearly improves the accuracy of small-sample estimation of *p*, but to what extent does it improve the accuracy of sample size estimation? To investigate this, I used true *p* from each problem-discovery database in Equation 1 to project the sample size required to achieve both 90% and 95% problem discovery for each database (true *n*, shown in Table 6). Next, I projected the required sample sizes using both the unadjusted and the GT-adjusted estimate of *p* at each level of sample size (2–6). The final step was to cal-

Table 6: Projected Sample Size Requirements for Each Problem-Discovery Database

Database	True p^a	90% Problem Discovery		95% Problem Discovery	
		n	*Proportion*[b]	*n*	*Proportion*[b]
MACERR[c]	.16	14	.913	18	.957
VIRZI90[d]	.36	6	.931	7	.956
MANTEL[e]	.38	5	.905	7	.963
SAVINGS[e]	.26	8	.906	11	.961

[a]True *p* for the specified database. [b]Proportion of problems discovered at that sample size. [c]Lewis (1994) and Lewis, Henry, and Mack (1990). [d]Virzi (1990, 1992). [e]Nielsen and Molich (1990).

Table 7: Deviation From Required Sample Size for No Adjustment (NONE) and Good–Turing (GT) Discounting

Sample Size (N)	NONE 90	GT 90	NONE 95	GT 95
2	5.5	2.8	7.3	4.0
3	4.5	1.8	6.0	2.5
4	4.0	1.3	5.0	1.5
5	3.5	0.8	4.5	1.3
6	2.8	0.5	3.8	0.5

Note.   90 = 90% problem discovery; 95 = 95% problem discovery.

culate the difference between the values of true n and the unadjusted and adjusted estimates of n. These differences appear in Table 7, with positive values indicating underestimation of true n (the expected consequence of overestimating p).

An analysis of variance (ANOVA) on the underestimation data revealed significant main effects of sample size, $F(4, 12) = 21.4$, $p = .00002$ and discounting, $F(1, 3) = 15.3$, $p = .03$, and a significant Sample Size × Goal interaction (in which the goals are 90% and 95% problem discovery), $F(4, 12) = 3.6$, $p = .04$. As the size of the sample used to estimate p increased, the magnitude of underestimation in the projected sample size decreased. GT estimation reduced the magnitude of underestimation relative to no adjustment. Although the underestimation for 95% discovery consistently exceeded that for 90% discovery, as the sample size used to estimate p increased, the difference between the magnitude of underestimation for 90% and 95% discovery decreased.

Discussion.   Adjustment of p using discount methods provided a much more accurate estimate of the true value of p than unadjusted estimation. The best known of the methods, GT discounting, was as effective as any of the other evaluated discounting methods. GT estimation, though, still generally left the estimate of p slightly inflated, leading to some underestimation of required total sample sizes when projecting from the initial sample. The magnitude of this underestimation of required sample size decreased as the size of the initial sample used to estimate p increased. For initial sample sizes of 4 to 6 participants, the magnitude of underestimation ranged from about 1.5 to 0.5 participants. Thus, final sample sizes projected from GT estimates based on initial sample sizes of 6 participants should generally be quite accurate. For each of the investigated problem-discovery databases and both problem-discovery goals, the mean extent of underestimation of the required sample size never exceeded 1 participant when estimating p from a 6-participant sample. (For a more comprehensive analysis of this data, see Lewis, 2000d.)

3.3. Development of Regression Equations for Predicting True p

The purpose of the third Monte Carlo experiment was to develop linear regression equations that estimate the true value of p using data available from a problem-discovery usability study.

Source data.   To obtain data for the generation of the regression equations, I divided the errors in the MACERR database into four groups with mean p of .10, .25, .50, and .73. The use of these four groups ensured the presence of training data for the regression equations with a range of values for true p. A Monte Carlo procedure generated 1,000 cases for each group for each level of sample size from 2 to 6 (20,000 cases). Each case included the following measurements:

- Unadjusted estimate of p.
- Normalized estimate of p.
- Sample size.
- True value of p.

Regression equations.   I used SYSTAT (Version 5) to create three simple regression models (predicting true p with the initial estimate of p only, with the normalized estimate of p only, and with the sample size only) and three multiple regression models (predicting true p with a combination of the initial estimate of p and the sample size, the normalized estimate of p and the sample size, and both the initial and normalized estimates of p and the sample size). Table 8 contains the resulting regression equations, the percentage of variance explained by the regression (R^2) and the observed significance level of the regression (osl).

In Table 8, *truep* is the true value of p as predicted by the equation, *estp* is the unadjusted estimate of p from the sample, *normp* is the normalized estimate of p from the sample, and n is the sample size used to estimate p. All regressions (REG) except for REG3 (using only n) were significant. For all significant regressions, t tests for the elements of the equations (constants and beta weights) were all statistically significant ($p < .0001$). The percentage of explained variance was highest for REG2, REG4, REG5, and REG6. Because the previous Monte Carlo studies indicated that the sample size plays an important role when estimating p, REG4, REG5, and REG6 received further evaluation in the following experiment.

Table 8: Regression Equations for Predicting True p

Key	Equation	R^2	osl
REG1	$truep = -.109 + 1.017 {*} estp$	.699	0.000
REG2	$truep = .16 + .823 {*} normp$	.785	0.000
REG3	$truep = .396 + 0 {*} n$	.000	1.000
REG4	$truep = -.387 + 1.145 {*} estp + .054 {*} n$	.786	0.000
REG5	$truep = .210 + .829 {*} normp - .013 {*} n$	.791	0.000
REG6	$truep = -.064 + .520 {*} estp + .463 {*} normp + .017 {*} n$	.799	0.000

Note. R^2 = percentage of variance explained by the regression; osl = observed significance level of the regression; REG = regression; *truep* = true value of p as predicted by the equation; *estp* = unadjusted estimate of p from the sample; *normp* = normalized estimate of p from the sample; n = sample size used to estimate p.

3.4. Evaluation of Normalization and Regression

The purpose of the fourth Monte Carlo experiment was to evaluate and compare unadjusted accuracy and the accuracy of adjustments using the normalization procedure (described previously in Section 2.1), regression equations (REG4, REG5, and REG6), and the GT procedure.

Evaluation of accuracy.   An ANOVA on the *RMSE* data indicated significant main effects of sample size, $F(4, 12) = 84.0$, $p = .00000009$ and adjustment method, $F(5, 15) = 32.7$, $p = .0000002$, and a significant interaction between these effects, $F(20, 60) = 20.6$, $p = .000008$. Figure 3 illustrates the interaction.

Figure 3 shows that as the sample size increased, accuracy generally increased for all estimation procedures (the main effect of sample size). A set of planned t tests evaluated, at each level of sample size, the significance of difference between no adjustment and each adjustment method and the significance of difference between GT estimation (the discounting method selected for further evaluation) and the other procedures. For sample sizes of 2, 3, and 4, all adjustment methods improved estimation accuracy of p relative to no adjustment (15 tests with 3 *df* each, all $ps < .02$). At a sample size of 5, the accuracy of REG6 was no longer significantly more accurate than no adjustment ($p > .10$). At a sample size of 6, both REG4 and REG6 (which included the unadjusted estimate of p as an element in the equation) failed to be more accurate than the unadjusted estimate

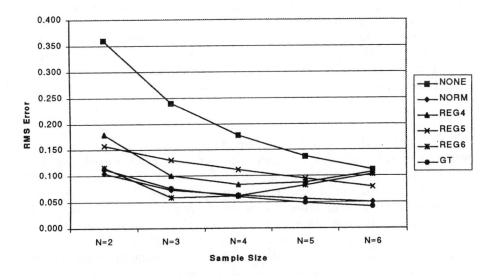

FIGURE 3   Adjustment Method × Sample Size interaction for root mean square error (RMSE).

**Table 9: Deviation From Required Sample Size as a Function
of Adjustment Method, Sample Size, and Discovery Goal**

Sample Size (N)	NONE 90	GT 90	NORM 90	NONE 95	GT 95	NORM 95
2	5.5	2.8	−0.8	7.3	4.0	−0.8
3	4.5	1.8	−0.8	6.0	2.8	−0.8
4	4.0	1.0	−0.8	5.0	1.5	−1.0
5	3.5	0.8	−1.3	4.5	1.0	−1.5
6	2.8	0.5	−1.3	3.8	0.5	−1.5

Note.   NONE = no adjustment; 90 = 90% problem discovery; GT = Good–Turing estimate; NORM = normalization estimate; 95 = 95% problem discovery.

(both $ps > .60$). Overall, GT and normalization produced the most accurate adjustments in this evaluation.

Estimates of required sample size.   Using the same method as that used to evaluate sample-size projections based on GT adjustment, Table 9 shows the difference between the values of true n (see Table 6) and the unadjusted and adjusted (both GT and normalization) estimates of n. Positive values indicate underestimation of true n; negative values indicate overestimation.

An ANOVA revealed significant main effects of sample size, $F(4, 12) = 6.8$, $p = .004$ and adjustment, $F(2, 6) = 4.7$, $p = .05$, and significant Sample Size × Problem-Discovery Goal, $F(4, 12) = 3.2$, $p = .05$ and Sample Size × Adjustment Method, $F(8, 24) = 3.6$, $p = .007$ interactions. As the size of the sample used to estimate p increased, the magnitude of deviation from true n in the projected sample size decreased. Both GT and normalized estimation reduced the magnitude of deviation relative to no adjustment; but this magnitude decreased as the sample size increased. Although the deviation for 95% discovery consistently exceeded that for 90% discovery, as the sample size increased, the difference between the magnitude of deviation for 90% and 95% discovery decreased.

Discussion.   The attempts to develop multiple regression equations for the prediction of true p did not fare as well as the nonregression approaches of GT and normalization. Even if the regression-based approaches had been as accurate (as measured by *RMSE*) as the nonregression approaches, the nonregression approaches would be preferable because they do not rely on statistically estimated parameters, making them solutions that have potentially greater generalizability. (For a more comprehensive analysis of this data, see Lewis, 2000c.)

As in the previous evaluation, adjustment with GT discounting consistently resulted in a slight underestimation of the required sample size. Adjustment via normalization consistently resulted in a slight overestimation of the required sample size. This result suggests the intriguing possibility that adjustment based on a com-

bination of GT and normalization procedures might yield highly accurate estimates of true p and n.

It is possible that nonlinear regression might have yielded superior results to linear regression. Exploring such alternatives, however, increases the potential for
capitalizing on chance patterns in the data. If the combination of GT and normalization were to yield the expected estimation accuracy, then there would be no need
to investigate solutions based on more complex regression models.

3.5. Improved Estimation Through the Combination of GT and Normalization

The purpose of this experiment was to assess the improvement in accuracy regarding the estimation of true p obtained by combining estimates of p calculated
with the normalization and GT methods. This is important because even though
the results of the previous experiment indicated that combining GT and normalization should work well, that judgment came from the averaging of average estimates—something that would be impossible for a practitioner conducting a single usability study. To control for this in this experiment, the program computed
the combination estimate for each case generated via Monte Carlo simulation.
Doing the computation at this level made it possible to evaluate the properties of
the distribution of the combined estimate (which was not possible given the data
in the previous experiment).

The adjustment methods explored in this experiment were no adjustment
(NONE), GT discounting, normalization (NORM), and the combination
(through simple averaging) of GT and NORM (COMB). The formula for this
combination was:

$$truep = \frac{1}{2}[(estp - 1/n)(1 - 1/n)] + \frac{1}{2}[estp/(1 + GTadj)] \qquad (2)$$

where $truep$ is the adjusted estimate of p calculated from the estimate of p derived
from the participant by problem matrix ($estp$), n is the sample size used to compute
the initial estimate of p, and $GTadj$ is the GT adjustment to probability space, which
is the proportion of the number of problems that occurred once divided by the number of different problems (see Section 2.1).

Estimates of RMSE.   An ANOVA on the $RMSE$ data indicated significant
main effects of sample size, $F(4, 12) = 94.3$, $p = .00000003$ and adjustment method,
$F(3, 9) = 64.4$, $p = .000002$, and a significant interaction between these effects, $F(12, 36) = 36.5$, $p = .0000002$. Figure 4 illustrates the interaction.

Figure 4 shows that as the sample size increased, accuracy generally increased
for all estimation procedures (the main effect of sample size). The lines for NORM,
GT, and COMB almost overlaid one another, with COMB having slightly less

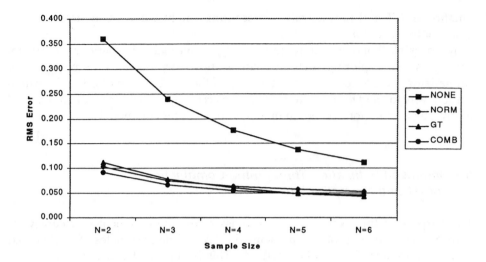

FIGURE 4   Root mean square error (RMSE) as a function of sample size and adjustment method.

RMSE than either GT or NORM. A set of planned *t* tests showed that estimates based on this combination resulted in significantly lower *RMSE* than unadjusted estimates for all sample sizes (all *p*s < .02). A similar set of *t* tests showed that none of the *RMSE* differences among GT, normalization, or their combination were significant (all *p*s > .10). In this analysis, the source of the significance of the main effect of adjustment type was solely due to the difference between unadjusted and adjusted estimates of *p*.

Estimates of required sample size.   I conducted an ANOVA on the deviations from required sample size for unadjusted *p*, GT discounting, normalization, and the averaging of GT and normalization, treating databases as subjects in a within-subjects design with independent variables of sample size, type of adjustment, and discovery goal (with levels of 90% and 95%). The main effects of sample size, $F(4, 12) = 5.6, p = .009$ and adjustment, $F(3, 9) = 4.9, p = .03$ were significant, as were the Discovery Goal × Adjustment Type interaction, $F(3, 9) = 3.9, p = .05$ and the Sample Size × Adjustment Type interaction, $F(12, 36) = 2.9, p = .006$. In the Discovery Goal × Adjustment Type interaction, the underestimation of the required sample size for 95% was generally greater than for 90%, except for the normalization adjustment type, which had equal deviation for both levels of discovery goal (see Figure 5). The Sample Size × Adjustment Type interaction indicated a general decline in the magnitude of underestimation as a function of the sample size used to estimate *p*. This trend seemed strong for estimates based on unadjusted *p*, GT estimates, and the combination estimate, but not for estimates based on normalized *p* (see Figure 6).

As expected, across all sample sizes the GT estimate tended to underestimate the required sample size and the normalized estimate tended to overestimate the re-

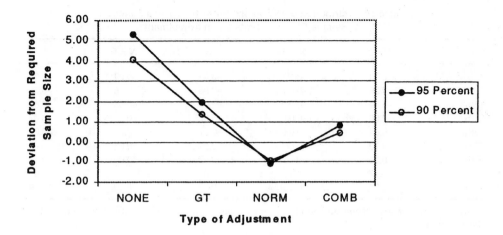

FIGURE 5   Discovery Goal × Adjustment Method interaction.

quired sample size. For the combination estimate of p based on sample sizes of 4, 5, and 6 participants, the estimates of required sample sizes had almost no deviation from true n. For estimates of p computed from initial sample sizes of 2 and 3 participants, the mean underestimations of true n projected from combination-adjusted estimates of p were 2 participants and 1 participant, respectively.

Variability of combination estimate.   The preceding analyses focused on the means of various distributions. The analyses in this section address the distribution of p after adjustment with the normalization and GT combination procedure (spe-

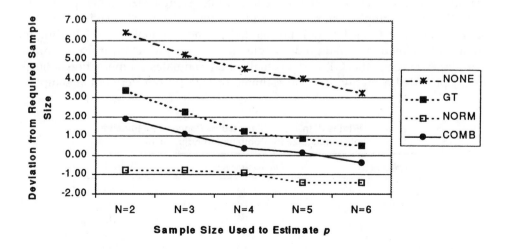

FIGURE 6   Sample Size × Adjustment Method interaction.

**Table 10: Distribution of Deviations From True *p* as a Function
of Sample Size for Combination Adjustment**

Percentile	N = 2	N = 3	N = 4	N = 5	N = 6
1st	−0.10	−0.09	−0.09	−0.09	−0.08
5th	−0.06	−0.06	−0.06	−0.06	−0.06
10th	−0.04	−0.04	−0.05	−0.05	−0.05
25th	0.00	−0.01	−0.02	−0.03	−0.03
50th	0.04	0.02	0.01	0.00	−0.01
75th	0.10	0.06	0.04	0.03	0.02
90th	0.14	0.10	0.07	0.05	0.04
95th	0.17	0.12	0.09	0.07	0.06
99th	0.24	0.16	0.12	0.10	0.09

Note. The 25th and 75th percentiles define the interquartile range. The 5th and 95th percentiles define the 90% range

cifically, the deviation of this adjustment from true *p*). These analyses will help practitioners understand the variability of this distribution so they can take this variability into account when planning problem-discovery usability studies.

Table 10 provides the average deviations from true *p* (collapsed across databases) as a function of sample size for adjustments using Equation 2 (the combination adjustment). Table 11 shows the corresponding magnitude of interquartile and 90% ranges for these deviations. Relatively smaller deviations from true *p* and relatively smaller ranges are indicators of better accuracy. The results shown in the tables indicate that increasing the sample size used to estimate *p* decreased both the magnitude and variability of deviation from true *p*.

Discussion. GT estimation generally left the estimate of *p* slightly inflated, leading to some underestimation of required total sample sizes when projecting from the initial sample. Estimating *p* with the normalization procedure had the same accuracy as GT estimation but tended to underestimate true *p*, leading to some overestimation of required sample sizes when projecting from an initial sample. Averaging the GT and normalization estimates (combination adjustment) provided a highly accurate estimate of true *p* from very small samples,

**Table 11: Widths of Interquartile and 90% Ranges as a Function
of Sample Size for Combination Adjustment**

Sample Size (N)	Interquartile Range	90% Range
2	.10	.23
3	.07	.18
4	.06	.15
5	.05	.13
6	.05	.13

which in turn led to highly accurate estimates of required sample sizes for specified problem-discovery goals. These estimates appear to be accurate enough that a practitioner should be able to make an initial projection from a combination-adjusted estimate of p using a sample with as few as 2 participants and will generally not underestimate the required sample size by much. A more conservative approach would be to use the normalized estimate of p when projecting from sample sizes with 2 or 3 participants (which should generally overestimate the required sample size slightly). The increased variation of p when estimated with a small sample also supports the use of the conservative approach. Using these techniques, usability practitioners can adjust small sample estimates of p when planning usability studies. As a study continues, practitioners can reestimate p and project the revised sample size requirement. (For a more comprehensive analysis of this data, see Lewis, 2000c.)

The apparent trends in Figure 6 indicated that it might not be wise to use the combination or normalization approaches when the sample size exceeds 6 participants. At 6 participants, normalization continued to underestimate p, and the combination approach began to slightly underestimate p. The GT approach appeared to be getting closer to true p and, as the sample size continues to increase, the unadjusted estimate of p should continue to approach true p. It was not clear from the data at what sample size a practitioner should abandon GT and move to the unadjusted estimate of p.

3.6. Evaluation of Best Procedures for Sample Sizes From 2 to 10

One goal of the final Monte Carlo experiment was to replicate the previous investigation of a variety of approaches (NORM, REG2, REG5, GT, COMB) for adjusting observed estimates of p to bring them closer to true p using sample sizes from 2 to 10 participants for the initial estimate of p. In particular, would the combination approach continue to provide accurate estimates of true p for sample sizes from 7 to 10 participants?

Another goal was to investigate the extent to which inaccuracy in estimating problem-discovery sample size using these methods affects the true proportion of discovered problems. The previous investigations assessed the deviation from the sample size required to achieve 90% and 95% problem-discovery goals but did not assess the magnitude of deviation from the problem-discovery goals caused by overestimating or underestimating the required sample size.

Estimates of p.   An ANOVA (within-subjects ANOVA treating problem-discovery databases as subjects), conducted on the problem-discovery rates (p) for each of the six estimation methods at each of the nine levels of sample size, revealed a significant main effect for type of adjustment, $F(8, 24) = 92.6, p = .00000015$; a significant main effect of sample size, $F(5, 15) = 138.5, p = .00000015$; and a significant Ad-

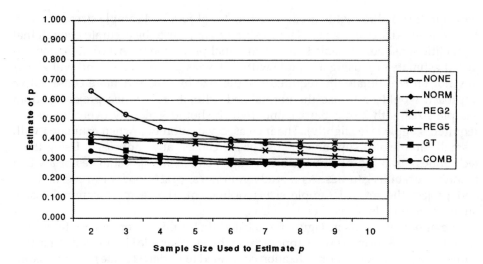

FIGURE 7   Adjustment Method × Sample Size interaction for problem-discovery rate.

justment Type × Sample Size interaction, $F(40, 120) = 72.8$, $p = .0001$. Figure 7 illustrates the interaction.

Estimates of RMSE.   An ANOVA conducted on *RMSE* revealed a significant main effect for type of adjustment, $F(8, 24) = 120.6$, $p = .00000004$; a significant main effect of sample size, $F(5, 15) = 27.2$, $p = .0000006$; and a significant Adjustment Type × Sample Size interaction, $F(40, 120) = 32.9$, $p = .0000001$ (see Figure 8).

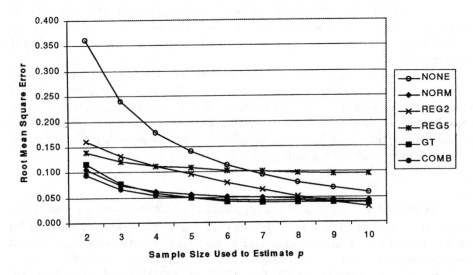

FIGURE 8   Adjustment Method × Sample Size interaction for root mean square error.

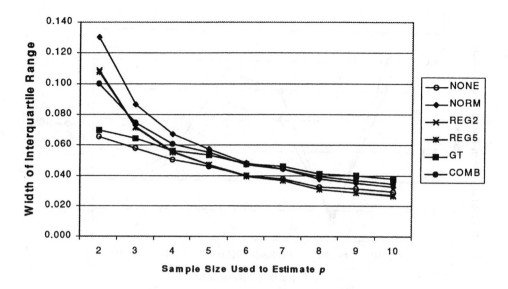

FIGURE 9   Adjustment Method × Sample Size interaction for interquartile range.

Estimation variability.   The *interquartile range* is the size of the interval that contains the central 50% of a distribution (the range from the 25th to the 75th percentile). The smaller this range, the less variable is the distribution. An ANOVA conducted on interquartile ranges revealed a significant main effect for type of adjustment, $F(8, 24) = 109.1, p = .0000003$; a significant main effect of sample size, $F(5, 15) = 10.3, p = .0002$; and a significant Adjustment Type × Sample Size interaction, $F(40, 120) = 69.8, p = .00000001$. Figure 9 illustrates this interaction.

Estimates of required sample size.   I conducted a within-subjects ANOVA on the deviations from required sample sizes, treating problem-discovery databases as subjects. The independent variables were adjustment method (NONE, NORM, REG2, REG5, GT, COMB), sample size used to estimate p (2–10), and problem-discovery goal (90%, 95%). The analysis indicated the following significant effects:

- Main effect of adjustment method, $F(5, 15) = 4.1, p = .015$.
- Main effect of sample size, $F(8, 24) = 3.8, p = .005$.
- Adjustment Method × Discovery Goal interaction, $F(5, 15) = 3.9, p = .019$.
- Adjustment Method × Sample Size interaction, $F(40, 120) = 3.2, p = .0000006$.
- Adjustment Method × Sample Size × Problem-Discovery Goal interaction, $F(40, 120) = 2.0, p = .002$.

Collapsed over discovery goal and sample size, the deviations from true n for the various methods of adjustment (main effect of adjustment method) were underestimations of 3.7 for NONE, 3.1 for REG5, 2.5 for REG2, 0.7 for GT, and overestimations of 1.3 for NORM and 0.1 for COMBO. The pattern for the main ef-

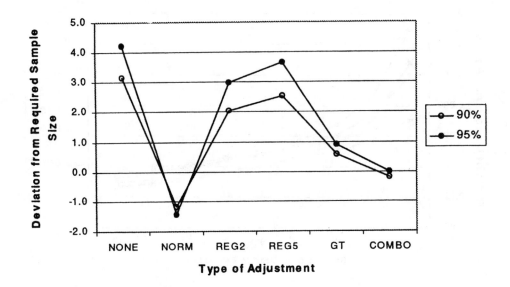

FIGURE 10   Deviation from required sample size as a function of adjustment method and discovery goal.

fect of sample size was a steady decline in overestimation from 3.1 at a sample size of 2 to 0.3 at a sample size of 10.

Figures 10 and 11 show the Adjustment Method × Discovery Goal interaction and the Adjustment Method × Sample Size interaction. The pattern for the Adjustment Method × Sample Size × Discovery Goal interaction was very similar to that

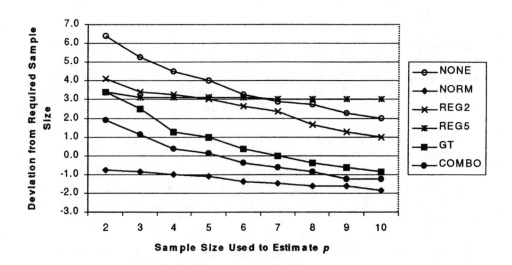

FIGURE 11   Deviation from required sample size as a function of adjustment method and sample size.

for the Adjustment Method × Sample Size interaction, with the difference between deviations for 90% and 95% discovery declining as a function of sample size. (Positive values indicate underestimation of the required sample size; negative values indicate overestimation.)

Deviation from problem-discovery goal.

Because the fundamental problem in underestimating a required sample size is the failure to achieve a specified problem-discovery goal, I also conducted a within-subjects ANOVA on the deviations from specified problem-discovery goals of 90% and 95% discovery, treating problem-discovery databases as subjects. The independent variables were adjustment method (NONE, NORM, REG2, REG5, GT, COMB), sample size used to estimate p (2–10), and problem-discovery goal (90%, 95%). The dependent variable was the difference between the specified problem-discovery goal and the magnitude of problem discovery for the projected sample sizes calculated in the previous section. Because sample sizes are discrete rather than continuous variables, there are cases in which a sample size that is 1 participant smaller than the sample size required to achieve (meet or exceed) a specified problem-discovery goal is actually much closer to the goal (although just under it) than the required sample size. In other words, underestimating the sample size requirement for a specified problem-discovery goal might, in many cases, still allow a practitioner to come very close to achieving the specified goal.

The analysis indicated the following significant effects:

- Main effect of adjustment method, $F(5, 15) = 13.6, p = .00004$.
- Main effect of sample size, $F(8, 24) = 15.5, p = .0000001$.
- Adjustment Method × Sample Size interaction, $F(40, 120) = 12.5, p = .005$.
- Sample Size × Problem-Discovery Goal interaction, $F(8, 24) = 3.8, p = .006$.

The pattern for the main effect of adjustment method was an underestimation of .113 for no adjustment, underestimation of .068 for REG5, underestimation of .053 for REG2, overestimation of .020 for NORM, underestimation of .008 for GT, and overestimation of .008 for COMBO. The pattern for the main effect of sample size was a steady decline in the magnitude of underestimation from .100 for a sample size of 2 to .008 for a sample size of 10. The basis for the Sample Size × Discovery Goal interaction was that deviation tended to be smaller for 90% discovery for sample sizes smaller than 6, with the opposite trend for sample sizes greater than 6. Figure 12 illustrates the interaction between adjustment method and sample size.

Iterative sample size estimation strategy using the combination adjustment.

One practical application for the use of the combined estimator is to allow usability practitioners to estimate their final sample size requirement from their first few participants. To do this, practitioners must keep a careful record of which

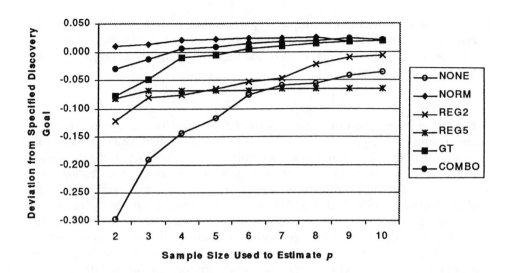

FIGURE 12 Deviation from specified discovery goal as a function of adjustment method and sample size.

participants experience which usability problems to enable calculation of an unadjusted initial estimate of p that they then adjust using the combined estimator.

Suppose, for example, that a practitioner is conducting a study on a product with problem-discovery characteristics similar to MACERR (true p of .16), and the practitioner has set a goal of discovering 90% of the problems. Referring to Table 5, the initial unadjusted estimate of p calculated from the first 2 participants would be, on average, .568. Adjusting this initial estimate with the combined procedure results in a value for p of .218. Using Equation 1 to project the sample size until the estimated proportion of discovery exceeds .900 yields a preliminary sample size requirement of 10 participants. After the practitioner runs 2 more participants toward the goal of 10 participants, he or she can recalculate the adjusted value of p to be .165 and can project the final required sample size to be 13 participants. As shown in Table 6, the sample size actually required to exceed a problem-discovery proportion of .900 is 14 participants. With 13 participants, the true proportion of problem discovery is .896, which misses the specified problem-discovery target goal by only .004—less than 0.5%.

Table 12 shows the outcome of repeating this exercise for each database and for each investigated problem-discovery goal. These outcomes show that following this procedure leads to very accurate estimates of sample sizes and very little deviation from problem-discovery goals—a remarkable outcome given the differences in the usability studies that produced these problem-discovery databases. The mean deviation from problem-discovery goal across cases was .006—overachievement by just over 0.5%.

A similar evaluation conducted for lower, less-aggressive problem-discovery goals (70%, 75%, and 80%) revealed a similar pattern of results. The overall mean

Table 12: Problem-Discovery Outcomes Achieved by Using Combination Adjustment to Project a Final Sample Size from Initial Sample Sizes of 2 and 4 Participants

Database	True p	Goal %	Est p | N = 2	Comb p | N = 2	n | N = 2	Est p | N = 4	Comb p | N = 4	n | N = 4	True n	Deviation From Goal
MACERR[a]	.16	90	.566	.218	10	.346	.165	13	14	-.004
		95	.566	.218	13	.346	.165	17	18	-.002
VIRZI90[b]	.36	90	.662	.361	6	.485	.328	8	6	.031
		95	.662	.361	7	.485	.328	8	7	.021
MANTEL[c]	.38	90	.725	.462	4	.571	.429	5	5	.005
		95	.725	.462	5	.571	.429	6	7	-.010
SAVINGS[c]	.26	90	.629	.311	7	.442	.277	8	8	.006
		95	.629	.311	9	.442	.277	10	11	-.002

Note. True p = value of p estimated from the entire database; Est p | N = 2 = unadjusted estimate of p given a sample size of 2 participants; Comb p | N = 2 = combination-adjusted estimate of p given a sample size of 2; n | N = 2 = projected sample size requirement given the combination-adjusted estimate of p estimated from a sample size of 2; Est p | N = 4 = unadjusted estimate of p given a sample size of 4 participants; Comb p | N = 4 = combination-adjusted estimate of p given a sample size of 4; n | N = 4 = projected sample size requirement given the combination-adjusted estimate of p estimated from a sample size of 4; True n = sample size requirement projected from True p.

[a]Lewis (1994) and Lewis, Henry, and Mack (1990). [b]Virzi (1990, 1992). [c]Nielsen and Molich (1990).

deviation from the discovery goal was overachievement of .021 (2.1%). The greatest mean deviation across the problem-discovery goals occurred for the VIRZI90 database (overachievement of .081), with smaller deviations for the other three databases (.000, .006, and −.004 for MACERR, MANTEL, and SAVINGS, respectively). Averaging over databases indicated overachievement of .047 and .028 for 70% and 75% discovery, and underachievement of .012 for 80% discovery. In all cases, the projected sample size given, $n = 4$, was within one participant of true n, with an overall average deviation of estimated sample size from true sample size requirement of −0.17 participants.

3.7. Discussion

The decision about which adjustment procedure or procedures to use should take into account both the central tendency and variability of distributions created by applying the adjustment procedure or procedures. A measure that produces mean estimates close in value to true p will generally be more accurate in the long run. A measure with low variability is less likely to produce an extreme outlier in any single study. Usability practitioners do not typically conduct large-scale studies, however, so it is important to balance benefits associated with the statistical long run with the benefits associated with reducing the risk of encountering an extreme outlier.

The regression equations (REG2 and REG5) tended to be less variable than other adjustment procedures, but their accuracy was very poor relative to all other adjustment procedures. The accuracy of REG2 improved as a function of the sample size used to estimate p, but the accuracy of REG5 did not. Their relatively poor performance removes these regression equations from consideration as a recommended method for adjusting p.

The remaining procedures (NORM, GT, and COMB) showed similar patterns to one another for deviations from true p. When estimating p with smaller sample sizes (2–4 participants), the curves for these three measures showed some separation, with the differences diminishing and the curves converging as the size of the sample used to estimate p increased. For these measures, especially at small sample sizes, the normalization procedure appeared to produce the best results and the combined estimator produced the second-best results.

The measures of interquartile range, however, indicated that the estimates produced by the normalization procedure were much more variable than those produced by the GT or the combined estimator. The results for the *RMSE* (which take both central-tendency accuracy and variability into account) showed that all three measures had essentially equal accuracy at all levels of sample size from 2 to 10. As expected, the variability of all measures decreased as a function of the sample size used to estimate p.

Considering all the information, the combined estimator seemed to provide the best balance between central-tendency accuracy and lower variability, making it the preferred adjustment procedure. What really matters, though, is the extent to which the adjustment procedure leads to accurate sample size estimation and achievement of specified problem-discovery goals.

The analyses of underestimation of required sample sizes and deviation from problem-discovery goals also support the use of the combined estimator. After averaging across all problem-discovery databases (MACERR, VIRZI90, MANTEL, and SAVINGS), both problem-discovery goals (90%, 95%), and sample sizes from 2 to 10, the accuracy of sample size estimation with the combined estimator was almost perfect, overestimating the required sample size by only 0.1 participant on average. The results were similar for mean deviation from problem-discovery goal. The magnitude of deviation was about the same for the combined estimator and the GT estimator. On average, however, the combined estimator tended to slightly overachieve the discovery goal, whereas the GT estimator tended to slightly underachieve the discovery goal.

With one exception, the patterns of results for the significant interactions supported the unqualified use of the combined estimator. The exception was the interaction between adjustment method and the sample size used to estimate p. As shown in Figure 11, at sample sizes of 2 and 3 the combined estimator tended to underestimate the required sample size, but the normalization procedure tended to overestimate it. The same interaction for the deviation from the discovery goal, however, indicates that the consequence of this underestimation of the required sample size was slight, even when the sample size used to estimate p was only 2 participants (in which case the underachievement was, on average, 3%). This does suggest some need on the part of practitioners to balance the cost of additional participants against their need to achieve a specific problem-discovery goal. If the former is more important, then the practitioner should use the combined estimator. If the latter is more important and the practitioner is estimating p once (not iteratively) from a very small sample size, then it would be reasonable to use the more conservative normalization procedure. (For a more comprehensive analysis of this data, see Lewis, 2000a.)

3.8. Summary of Analyses and Results

A series of Monte Carlo simulations provided evidence that the average of a normalization procedure and GT discounting produces highly accurate estimates of usability problem-discovery rates from small sample sizes. The motivation for conducting the research was the observation that unadjusted estimates of p derived from small-sample usability studies have a bias in a direction that would lead usability practitioners to believe that their studies have been more effective than the data really warrants (Hertzum & Jacobsen, this issue).

The simulations allowed investigation of three broad classes of methods for reducing initial estimates of p from small-sample usability studies: discounting procedures, regression, and normalization. Accuracy assessments of the various procedures (using *RMSE* as the measure of accuracy) indicated that (a) GT discounting (the best known of the discounting methods investigated) was as accurate as or more accurate than the other discounting procedures, (b) normalization was also a very accurate adjustment procedure, and (c) regression did not produce very satisfactory adjustments.

An unexpected (but fortuitous) result was that two of the most accurate adjustment procedures—GT discounting and normalization—had residual biases in opposite directions and of about equal magnitude. Investigation of an adjustment procedure based on the combination of these methods indicated that this approach provided the most satisfactory adjustments (high accuracy and low variability). Using this combined procedure to adjust initial estimates of p (derived from several published large-sample usability studies) resulted in highly accurate estimates of required sample sizes.

4. CONCLUSIONS AND RECOMMENDATIONS FOR PRACTITIONERS

- The overestimation of p from small-sample usability studies is a real problem with potentially troubling consequences for usability practitioners.
- It is possible to compensate for the overestimation bias of p calculated from small-sample usability studies.
- The combined normalization and GT estimator (Equation 2) is the best procedure for adjusting initial estimates of p calculated from small samples (2 to 10 participants).
- If (a) the cost of additional participants is low, (b) the sample size used to estimate p is very small (2 or 3 participants), and (c) it is very important to achieve or exceed specified problem-discovery goals, then practitioners should use the normalization procedure to adjust the initial estimate of p.
- Practitioners can obtain accurate sample size estimates for problem-discovery goals ranging from 70% to 95% by making an initial estimate of the required sample size after running 2 participants, then adjusting the estimate after obtaining data from another 2 (total of 4) participants.

REFERENCES

Chapanis, A. (1988). Some generalizations about generalization. *Human Factors, 30,* 253–267.
Cliff, N. (1987). *Analyzing multivariate data.* San Diego, CA: Harcourt Brace.
Connell, I. W., & Hammond, N. V. (1999). Comparing usability evaluation principles with heuristics: Problem instances vs. problem types. In M. A. Sasse & C. Johnson (Eds.), *Proceedings of INTERACT '99—Human–Computer Interaction* (Vol. 1, pp. 621–629). Edinburgh, Scotland: International Federation for Information Processing.
Draper, N. R., & Smith, H. (1966). *Applied regression analysis.* New York: Wiley.
Jelinek, F. (1997). *Statistical methods for speech recognition.* Cambridge, MA: MIT Press.
Lewis, J. R. (1982). Testing small-system customer set-up. *Proceedings of the Human Factors Society 26th Annual Meeting* (pp. 718–720). Santa Monica, CA: Human Factors Society.
Lewis, J. R. (1994). Sample sizes for usability studies: Additional considerations. *Human Factors, 36,* 368–378.

Lewis, J. R. (2000a). *Evaluation of problem discovery rate adjustment procedures for sample sizes from two to ten* (Tech. Rep. No. 29.3362). Raleigh, NC: IBM. Available from the author.

Lewis, J. R. (2000b). *Overestimation of p in problem discovery usability studies: How serious is the problem?* (Tech. Rep. No. 29.3358). Raleigh, NC: IBM. Available from the author.

Lewis, J. R. (2000c). *Reducing the overestimation of p in problem discovery usability studies: Normalization, regression, and a combination normalization/Good–Turing approach* (Tech. Rep. No. 29.3361). Raleigh, NC: IBM. Available from the author.

Lewis, J. R. (2000d). *Using discounting methods to reduce overestimation of p in problem discovery usability studies* (Tech. Rep. No. 29.3359). Raleigh, NC: IBM. Available from the author.

Lewis, J. R. (2000e). *Validation of Monte Carlo estimation of problem discovery likelihood* (Tech. Rep. No. 29.3357). Raleigh, NC: IBM. Available from the author.

Lewis, J. R., Henry, S. C., & Mack, R. L. (1990). Integrated office scenario benchmarks: A case study. In *Human Computer Interaction—INTERACT '90* (pp. 337–343). Cambridge, England: Elsevier, International Federation for Information Processing.

Manning, C. D., & Schutze, H. (1999). *Foundations of statistical natural language processing.* Cambridge, MA: MIT Press.

Nielsen, J. (1992). Finding usability problems through heuristic evaluation. In *Conference Proceedings on Human Factors in Computing Systems—CHI '92* (pp. 373–380). Monterey, CA: ACM.

Nielsen, J., & Landauer, T. K. (1993). A mathematical model of the finding of usability problems. In *Conference Proceedings on Human Factors in Computing Systems—CHI '93* (pp. 206–213). New York: ACM.

Nielsen, J., & Molich, R. (1990). Heuristic evaluation of user interfaces. In *Conference Proceedings on Human Factors in Computing Systems—CHI '90* (pp. 249–256). New York: ACM.

Norman, D. A. (1983). Design rules based on analyses of human error. *Communications of the ACM, 4,* 254–258.

Pedhazur, E. J. (1982). *Multiple regression in behavioral research: Explanation and prediction.* Fort Worth, TX: Harcourt Brace.

Virzi, R. A. (1990). Streamlining the design process: Running fewer subjects. In *Proceedings of the Human Factors Society 34th Annual Meeting* (pp. 291–294). Santa Monica, CA: Human Factors Society.

Virzi, R. A. (1992). Refining the test phase of usability evaluation: How many subjects is enough? *Human Factors, 34,* 443–451.

APPENDIX

MACERR Problem-Discovery Database

The MACERR database contains discovery information for 145 usability problems uncovered by observation of 15 participants, with p estimated to be .16 (Lewis, Henry, & Mack, 1990). The first column in the database table is the problem identification number. The next 15 columns represent the observation of the experience of each participant with that problem, with a 0 indicating that the participant did not experience the problem and a 1 indicating that the participant did experience that problem. The last column is the modal impact rating for the problem across participants experiencing the problem, using the behavioral rating scheme described in Lewis (1994). The criteria for the impact ratings were

Prob	P1	P2	P3	P4	P5	P6	P7	P8	P9	P10	P11	P12	P13	P14	P15	Impact
1	0	1	1	1	0	1	1	1	1	1	1	0	0	1	1	2
2	1	0	1	0	0	0	1	1	1	1	1	0	0	1	0	2
3	1	0	0	0	0	0	0	0	0	0	0	0	0	0	0	3
4	1	1	1	1	0	0	1	0	1	1	1	0	1	1	0	4
5	1	0	0	1	1	0	1	0	0	0	1	0	1	0	1	3
6	0	0	0	0	0	0	0	0	0	0	0	1	0	0	0	4
7	1	0	0	1	1	1	0	0	1	0	1	0	0	0	1	3
8	1	0	0	0	1	0	0	0	1	1	0	0	0	0	0	3
9	1	0	0	0	0	0	0	0	0	0	0	0	0	0	0	3
10	0	0	0	0	0	0	0	0	1	0	1	0	0	0	0	2
11	0	0	0	0	0	0	1	0	0	0	0	0	0	0	0	1
12	0	0	0	0	0	0	0	0	0	0	0	0	0	1	0	3
13	0	0	0	0	0	0	0	0	0	0	1	0	0	0	0	2
14	0	0	0	0	1	0	0	0	0	1	0	0	0	0	0	2
15	0	0	0	0	0	0	0	0	0	0	1	0	0	0	0	4
16	0	0	0	0	0	0	0	0	0	0	1	0	0	0	0	3
17	0	0	0	1	0	0	0	0	0	0	0	0	0	0	0	3
18	1	1	1	1	0	1	1	0	1	1	1	0	1	1	0	2
19	0	0	0	0	0	0	0	0	0	0	1	0	0	1	1	1
20	0	0	0	0	0	0	1	1	0	0	1	0	0	0	0	2
21	1	0	0	0	1	1	0	0	0	0	0	0	0	0	0	3
22	0	0	0	1	0	0	0	0	0	1	0	0	0	0	0	3
23	0	0	1	0	0	0	0	0	0	0	0	0	0	0	0	1
24	0	0	0	0	0	0	0	0	0	0	0	1	0	1	0	3
25	0	0	0	0	0	0	0	0	0	0	0	0	0	1	0	3
26	0	0	0	0	0	0	0	0	1	0	0	0	0	0	0	3
27	1	0	1	1	0	0	1	1	1	1	0	1	1	0	1	2
28	0	0	0	0	0	0	0	0	0	0	0	1	0	0	0	2
29	1	0	0	0	0	1	1	1	1	1	0	1	0	0	1	3
30	1	0	0	0	0	0	0	0	0	0	0	0	0	0	0	1
31	0	0	1	1	0	1	1	0	1	0	0	1	0	0	0	1
32	1	0	0	0	0	0	0	0	0	1	0	0	1	0	0	2
33	0	0	0	0	0	0	1	0	0	1	0	0	1	0	0	3
34	0	0	0	0	0	0	0	0	0	0	0	0	1	0	0	3
35	0	0	0	0	1	1	0	0	1	0	0	0	1	0	0	1
36	0	0	0	0	0	0	0	0	0	0	0	0	1	0	0	4
37	1	0	1	1	0	0	0	1	0	1	0	0	1	0	0	4
38	0	1	0	0	0	0	1	0	0	1	0	0	1	0	0	2
39	1	0	0	0	0	0	0	0	0	1	0	0	0	0	0	2
40	0	0	0	0	0	0	0	0	0	1	0	0	0	0	0	2
41	0	0	0	1	0	0	0	0	0	1	0	0	0	0	0	2
42	0	0	1	0	0	0	0	0	1	0	0	0	0	1	0	1
43	1	1	0	0	0	0	0	0	0	0	0	0	0	0	0	2
44	0	0	0	1	0	0	0	0	0	0	0	0	0	0	0	2
45	0	0	0	0	0	0	0	0	0	0	0	0	0	1	0	2
46	0	0	1	1	0	0	0	0	1	0	0	0	0	0	0	1
47	0	0	0	1	0	0	0	0	0	0	0	0	0	0	0	1
48	0	0	0	0	0	0	0	0	0	0	0	0	0	0	1	2
49	1	0	1	0	0	0	0	0	0	0	0	0	0	0	1	2
50	0	0	0	1	0	0	0	0	0	0	0	0	0	0	0	2

(continued)

Prob	P1	P2	P3	P4	P5	P6	P7	P8	P9	P10	P11	P12	P13	P14	P15	Impact
51	1	1	1	1	0	0	1	0	1	0	0	0	0	0	1	1
52	1	0	1	0	0	0	0	0	0	0	0	0	0	0	0	1
53	0	0	0	0	0	0	0	0	1	0	0	0	0	0	0	3
54	1	0	1	1	0	1	0	0	0	0	0	0	0	0	1	2
55	0	0	0	0	0	0	0	0	0	0	0	0	0	0	1	3
56	0	1	0	1	0	1	0	1	1	1	0	0	0	0	1	2
57	0	0	0	0	0	0	0	0	0	0	0	0	0	0	1	3
58	0	0	0	0	0	0	0	0	1	0	0	0	0	0	0	3
59	0	0	0	1	0	0	0	0	0	0	0	0	0	0	0	2
60	0	0	0	0	0	1	0	1	0	0	0	0	0	0	0	2
61	0	0	0	0	0	0	0	0	1	0	0	0	0	0	0	2
62	0	0	0	1	0	0	0	0	0	0	0	0	0	0	0	2
63	0	0	0	0	1	0	0	0	0	0	0	0	0	0	0	1
64	0	0	0	0	0	1	0	0	0	0	0	0	0	0	0	3
65	0	1	1	0	0	0	1	0	0	0	0	0	0	0	1	1
66	0	0	1	0	0	1	0	0	0	0	0	0	0	0	1	2
67	1	0	0	1	1	0	0	0	0	0	0	0	0	0	1	2
68	1	0	0	0	0	0	0	1	1	0	0	0	0	0	1	1
69	0	0	0	0	0	0	0	0	0	0	0	0	0	0	1	4
70	1	0	0	0	0	0	0	1	0	0	0	0	0	0	1	2
71	0	0	0	0	0	0	0	0	0	0	0	0	0	0	1	2
72	0	0	0	0	0	0	0	0	1	0	0	0	0	0	1	1
73	0	0	0	0	0	1	0	0	0	0	0	0	0	0	1	2
74	0	0	0	0	0	0	1	0	1	0	0	0	0	0	0	3
75	1	0	0	0	0	0	0	0	1	1	0	0	0	0	0	4
76	0	1	0	0	0	0	0	0	0	0	0	0	0	0	0	1
77	0	0	0	0	1	0	1	0	0	0	0	0	0	0	0	2
78	0	1	1	1	1	0	1	0	0	0	0	0	0	0	0	1
79	0	0	0	0	0	0	1	0	0	0	0	0	0	0	0	1
80	0	0	1	0	0	0	1	0	0	0	0	0	0	0	0	0
81	1	0	0	0	0	0	1	0	0	0	0	0	0	0	0	1
82	0	1	1	1	0	0	0	0	0	1	0	0	0	0	0	1
83	0	0	0	0	1	0	0	0	0	0	0	0	0	0	0	1
84	0	0	0	0	0	0	1	0	0	0	0	0	0	0	0	0
85	0	0	1	0	0	1	1	0	1	0	0	0	0	0	0	1
86	1	0	0	0	0	0	0	0	0	0	0	0	0	0	0	2
87	0	0	0	0	0	0	1	0	0	0	0	0	0	0	0	1
88	0	1	0	0	0	0	1	0	0	0	0	0	0	0	0	2
89	0	0	0	1	0	1	0	1	1	0	0	0	0	0	0	4
90	0	0	0	0	0	0	1	0	0	0	0	0	0	0	0	2
91	0	1	0	1	1	1	1	0	0	0	0	0	0	0	0	1
92	1	1	0	1	1	1	0	1	0	1	0	0	0	0	0	3
93	0	1	1	0	0	1	0	0	0	0	0	0	0	0	0	2
94	0	1	0	0	0	0	0	1	0	1	0	0	0	0	0	2
95	0	0	1	0	0	0	0	0	0	0	0	0	0	0	0	3
96	0	0	0	0	0	1	0	0	0	0	0	0	0	0	0	2
97	1	0	0	1	1	0	0	0	0	0	0	0	0	0	0	2
98	0	0	0	0	0	0	0	1	0	0	0	0	0	0	0	4
99	0	0	0	0	0	0	0	1	0	0	0	0	0	0	0	2
100	0	0	0	0	0	0	0	1	0	0	0	0	0	0	0	1

(continued)

Prob	P1	P2	P3	P4	P5	P6	P7	P8	P9	P10	P11	P12	P13	P14	P15	Impact
101	0	0	0	0	0	0	0	1	0	0	0	0	0	0	0	2
102	0	1	1	0	1	0	0	1	0	0	0	0	0	0	0	3
103	0	0	1	0	0	0	0	0	0	0	0	0	0	0	0	2
104	0	1	1	0	1	1	0	1	1	1	0	0	0	0	0	1
105	0	0	0	0	0	0	0	1	0	0	0	0	0	0	0	1
106	0	0	0	0	0	0	0	0	0	1	0	0	0	0	0	3
107	1	1	1	1	1	1	0	0	1	0	0	0	0	0	0	3
108	0	0	0	0	0	1	0	0	0	0	0	0	0	0	0	3
109	1	0	0	0	1	0	0	0	0	0	0	0	0	0	0	2
110	0	1	0	0	1	0	0	0	1	0	0	0	0	0	0	1
111	0	0	0	1	0	0	0	0	0	0	0	0	0	0	0	0
112	0	0	0	0	0	0	0	0	1	0	0	0	0	0	0	2
113	0	0	0	1	1	1	0	0	1	0	0	0	0	0	0	2
114	0	0	0	0	0	1	0	0	1	0	0	0	0	0	0	3
115	0	0	1	0	0	0	0	0	0	0	0	0	0	0	0	2
116	0	0	1	0	0	0	0	0	0	0	0	0	0	0	0	2
117	0	0	0	0	0	0	0	0	0	1	0	0	0	0	0	2
118	0	0	0	0	0	1	0	0	0	1	0	0	0	0	0	2
119	1	0	0	1	1	0	0	0	0	1	0	0	0	0	0	1
120	0	0	0	0	1	0	0	0	0	1	0	0	0	0	0	1
121	0	0	0	1	1	1	0	0	0	0	0	0	0	0	0	1
122	0	0	0	0	0	0	0	0	0	1	0	0	0	0	0	2
123	0	0	0	0	0	0	0	0	0	1	0	0	0	0	0	1
124	1	0	0	0	1	1	0	0	0	0	0	0	0	0	0	2
125	0	0	0	0	0	0	0	0	0	1	0	0	0	0	0	3
126	0	0	0	0	0	1	0	0	0	0	0	0	0	0	0	2
127	0	0	0	0	0	1	0	0	0	0	0	0	0	0	0	2
128	0	0	0	0	0	1	0	0	0	0	0	0	0	0	0	2
129	0	0	0	0	0	1	0	0	0	0	0	0	0	0	0	1
130	0	0	0	0	0	1	0	0	0	0	0	0	0	0	0	1
131	1	0	0	0	0	0	0	0	0	0	0	0	0	0	0	1
132	1	1	1	0	0	0	0	0	0	0	0	0	0	0	0	1
133	1	0	0	0	0	0	0	0	0	0	0	0	0	0	0	2
134	1	0	0	0	0	0	0	0	0	0	0	0	0	0	0	2
135	1	0	0	0	1	0	0	0	0	0	0	0	0	0	0	2
136	1	0	0	0	0	0	0	0	0	0	0	0	0	0	0	2
137	0	1	0	0	0	0	0	0	0	0	0	0	0	0	0	1
138	0	1	0	0	0	0	0	0	0	0	0	0	0	0	0	2
139	0	0	1	0	0	0	0	0	0	0	0	0	0	0	0	1
140	0	0	0	1	0	0	0	0	0	0	0	0	0	0	0	2
141	0	0	0	1	0	0	0	0	0	0	0	0	0	0	0	1
142	0	0	0	0	1	0	0	0	0	0	0	0	0	0	0	2
143	0	0	0	0	1	0	0	0	0	0	0	0	0	0	0	2
144	0	0	0	0	1	0	0	0	0	0	0	0	0	0	0	2
145	0	0	0	0	1	0	0	0	0	0	0	0	0	0	0	1

Note. Prob = problem identification number; P = participant; Impact = modal impact rating for the problem across participants; 0 = participant did not experience the problem; 1 = participant did experience the problem.

1. *Scenario failure.* Participants failed to successfully complete a scenario if they either requested help to complete it or produced an incorrect output (excluding minor typographical errors).
2. *Considerable recovery effort.* The participant either worked on error recovery for more than a minute or repeated the error within a scenario.
3. *Minor recovery effort.* The participant experienced the problem only once within a scenario and required less than a minute to recover.
4. *Inefficiency.* The participant worked toward the scenario's goal but deviated from the most efficient path.

In the analyses in this article I did not use the impact ratings, but I have provided them for any researcher who needs them. The database is available in electronic form in Lewis (2000e).

INTERNATIONAL JOURNAL OF HUMAN–COMPUTER INTERACTION, 13(4), 481–499
Copyright © 2001, Lawrence Erlbaum Associates, Inc.

The Effect of Perceived Hedonic Quality on Product Appealingness

Marc Hassenzahl

Department of Psychology, Darmstadt University of Technology
Germany

Usability can be broadly defined as quality of use. However, even this broad definition neglects the contribution of perceived fun and enjoyment to user satisfaction and preferences. Therefore, we recently suggested a model taking "hedonic quality" (HQ; i.e., non-task-oriented quality aspects such as innovativeness, originality, etc.) and the subjective nature of "appealingness" into account (Hassenzahl, Platz, Burmester, & Lehner, 2000).

In this study, I aimed to further elaborate and test this model. I assessed the user perceptions and evaluations of 3 different visual display units (screen types). The results replicate and qualify the key findings of Hassenzahl, Platz, et al. (2000) and lend further support to the model's notion of hedonic quality and its importance for subjective judgments of product appealingness.

1. INTRODUCTION

Since the 1980s, usability as a quality aspect of products has become more and more important. Nevertheless, the exact meaning of the term *usability* remains fuzzy. There are at least two distinct perspectives on usability (Bevan, 1995). It can either be thought of as a narrow product-oriented quality aspect complementing, for example, reliability (i.e., free of error) and portability, or as a broad, general "quality of use"; in other words, "that the product can be used for its intended purpose in the real world" (Bevan, 1995, p. 350).

The quality of use approach defines the usability of a product as its efficiency and effectiveness, together with the satisfaction of the user in a given "context of use" (see Bevan & Macleod, 1994; International Organization for Standardization

---

Marc Hassenzahl is now at Darmstadt University of Technology. I am grateful to Manfred Wäger and Silke Gosibat from Siemens Corporate Technology—User Interface Design (CT IC 7) for collecting the data presented in this article and giving me the opportunity to analyze and report it. I also thank Jim Lewis for helpful methodological and editorial comments and Uta Sailer for many clarifying comments on an earlier draft of this article.

Request for reprints should be sent to Marc Hassenzahl, Darmstadt University of Technology, Department of Psychology, Steubenplatz 12, 64293 Darmstadt, Germany. E-mail: hassenzahl@psychologie.tu-darmstadt.de

[ISO], 1998). Efficiency and effectiveness are product characteristics that can be objectively assessed, whereas satisfaction is the subjectively experienced positive or negative attitude toward a given product (ISO, 1998). Taking a close look at the actual measurement of satisfaction, it appears that some current approaches may test user's recognition of design objectives rather than actual user satisfaction. For example, the Software Usability Measurement Inventory (SUMI; Kirakowski & Corbett, 1993; see also Bevan & Macleod, 1994) contains five subscales (Efficiency, Helpfulness, Control, Learnability, and Positive Affect). Except for affect, all of these aspects map onto the efficiency or effectiveness claim of quality in use. To give a second example, the End User Computing Satisfaction Instrument (Doll & Torkzadeh, 1988; Harrison & Rainer, 1996) describes satisfaction as a higher order construct, including content and accuracy of the provided information, ease of use, timeliness, and format of information. Again, all these subdimensions point at the perceived efficiency or effectiveness of the product. Satisfaction, so this perspective assumes, is the mere consequence of recognizing the quality designed "into" a product, leading to a simple equation: If users perceive the product as effective and efficient, they will be satisfied. Thus, assuring efficiency and effectiveness and making it obvious to the user should guarantee satisfaction.

Studies from the Technology Acceptance literature suggest a more complex perspective. A study investigating the impact of perceived usefulness (i.e., usability and utility) and perceived fun on usage of software products, and user satisfaction in a work context (Igbaria, Schiffman, & Wieckowski, 1994) demonstrated an almost equal effect of perceived fun and perceived usefulness on system usage. Perceived fun had an even stronger effect on user satisfaction than perceived usefulness. The authors concluded that "fun features" (e.g., sounds, games, cartoons) might encourage people to work with new software products.

An experiment by Mundorf, Westin, and Dholakia (1993) partly confirmed this conclusion. They analyzed the effect of so-called hedonic components (i.e., nontask-related "fun factors" such as music) on enjoyment and intention of using a screen-based information system. They varied some of the most basic hedonic components: color versus monochrome, graphics versus nongraphics, and music versus nonmusic. The inclusion of hedonic components increased enjoyment as well as usage intentions. Hence, fun or enjoyment seems to be an aspect of user experience that contributes to overall satisfaction with a product. Moreover, fun or enjoyment may be stimulated by product features that do not necessarily increase user efficiency and effectiveness—or that even partially hamper those quality aspects (Carroll & Thomas, 1988). Thus, even the broad definition of quality of use as previously presented omits the important aspect of hedonic quality. This calls for an expanded concept of usability.

1.1. Appealing Products: A Suggested Research Model

Hassenzahl, Platz, et al. (2000) recently suggested and tested a research model that addresses ergonomic (usability) and hedonic aspects as key factors for appealing—and thus satisfying—products. It consists of three separate layers: (a) objective product quality (intended by the designers), (b) subjective quality perceptions and evaluations (cognitive appraisal by the users), and (c) behavioral and emotional consequences (for the user).

Hassenzahl, Platz, et al. (2000) suggest that a product might be described by a large number of different quality dimensions (e.g., predictability, controllability, etc.) grouped into two distinct quality aspects: ergonomic quality and hedonic quality. *Ergonomic quality* (EQ) refers to the usability of the product, which addresses the underlying human need for security and control. The more EQ a product has, the easier it is to reach task-related goals with effectiveness and efficiency. EQ focuses on goal-related functions or design issues. *Hedonic quality* (HQ) refers to quality dimensions with no obvious—or at least a second order—relation to task-related goals such as originality, innovativeness, and so forth. HQ is a quality aspect addressing human needs for novelty or change and social power (status) induced, for example, by visual design, sound design, novel interaction techniques, or novel functionality. A product can possess more or less of these two quality aspects.

To have an effect, users must first perceive these different quality aspects. A product intended to be very clear in presenting information inevitably fails if users cannot perceive this intended clarity. The correspondence of intended and apparent (perceived) quality of a product can be low (Kurosu & Kashimura, 1995), indicating differences in how designers (or usability experts) think of a product and how the users perceive it. If the major design goal is the efficiency or effectiveness of the product, objective usability is the appropriate evaluation target (e.g., operationalized by performance measures). By contrast, if the goal is to design a rich and appealing user experience (Laurel, 1993) with a product, focusing on individuals' perceptions (i.e., subjective usability, user-perceived quality) would be more appropriate (Leventhal et al., 1996). This justifies a separate model layer that explicitly addresses user perceptions.

On the basis of their perceptions, users may form a judgment of the product's appealingness (APPEAL). In other words, the user may weight and combine the perceived EQ and HQ into a single judgment. APPEAL (or a lack of) manifests itself in a global judgment about the products (e.g., good vs. bad).

Perception and its subsequent evaluation are similar to the concept of cognitive appraisal. This appraisal can have two different outcomes. On one hand, it may lead to behavioral consequences, such as increased usage frequency, increased quality of work results, or decreased learning time. On the other hand, it may lead to emotional consequences, such as feelings of enjoyment or fun or satisfaction (or frustration or distress or disappointment). To view emotions as a consequence of the user's cognitive appraisal is generally consistent with numerous emotion theories (see Ekman & Davidson, 1994, for an overview; Ortony, Clore, & Collins, 1988, for a more specific example). Behavioral and emotional consequences may be related to each other. For example, Igbaria et al. (1994) found that perceived fun and actual usage of software systems were correlated.

The proposed model puts emotions such as fun, satisfaction, joy, or anger slightly out of focus, viewing them as consequences of a cognitive appraisal process. The model's primary concern is determining whether HQ perceptions are a valuable road to increase a product's APPEAL. However, it is likely that a product that emphasizes HQ may elicit different emotions from a product that emphasizes EQ.

Figure 1 summarizes the key elements of the proposed research model.

In a study directed at testing parts of the research model, Hassenzahl, Platz, et al. (2000) concentrated on the cognitive appraisal stage of the model. Specifically, they investigated (a) whether users perceive EQ and HQ independently and (b) how the judgment of APPEAL depends on the perceptions of the two different quality aspects:

1. Robinson (1993, as cited in Leventhal et al., 1996) stated that the artifacts people choose to use are interpretable as statements in an ongoing "dialog" that people have with their environment. He suggested that the artifact's quality aspects, which satisfy the requirements of that dialog, may be independent from the artifact's usability and utility. Taking this assumption into account, one may expect independent perception of HQ and EQ. Indeed, a principal components analysis (PCA) of perceived quality dimensions in the form of a semantic differential (e.g., simple–complex, dull–exciting) revealed two independent components consistent with the a priori proposed EQ and HQ groups. Internal consistency (Cronbach's α) of both EQ and HQ proved to be very high (> .90). This confirmed the assumption that users perceive HQ and EQ consistently, distinguishing task-related aspects from non-task-related aspects.

2. It is possible that although individuals may perceive HQ, they do not consider it important for a judgment of APPEAL. In other words, users may find a product innovative (possessing HQ) but may be indifferent about it or find other aspects much more important. Hassenzahl, Platz, et al. (2000) measured APPEAL in the form of a semantic differential (e.g., good–bad; Cronbach's α = .95). To determine the contribution of HQ and EQ to APPEAL, they performed a regression analysis. The results showed an almost equal contribution of HQ and EQ to APPEAL. The

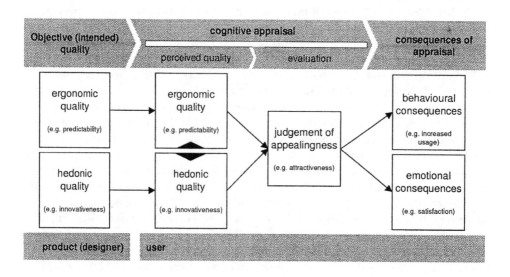

FIGURE 1   Research model.

data indicated a simple averaging model in which EQ and HQ have equal weights. This finding is consistent with the averaging model of information integration theory (Anderson, 1981), which proposes the integration of different pieces of information by an averaging process into a multiattributive judgment (such as APPEAL).

To conclude, users perceive EQ and HQ aspects independently, and these qualities appear to contribute equally to the overall judgment of a product's APPEAL. These findings support the cognitive appraisal layer of the research model.

1.2. Aims of This Study

Hassenzahl, Platz, et al. (2000) also discussed some limitations of their study: First, a lack of ecological validity is apparent. They used special stimuli (software prototypes) to induce EQ and HQ perceptions. The observed effects may heavily depend on the stimulus material provided, limiting generalizability. Furthermore, participants had to perceive and evaluate stimuli (software prototypes) that represented a very simple task from a domain that was not relevant for the participants' day-to-day life. In a sense, the whole study was more or less context free, which may have led to an artificial situation.

Second, the participants had an interaction time of only about 2 min with each stimulus (software prototype). This very short interaction time may have led to superficial cognitive processing of the stimuli, which in turn may have influenced the results.

Third, although internal consistency and factorial validity are important indicators of the reliability of the used semantic differential, the question of whether the scales really measure the hypothesized quality aspects and the user's evaluation (i.e., the construct validity of the scales) remains unanswered.

To provide further information about the validity of the research model for appealing products, I examine in this article some of those limitations and some additional aspects. Specifically, I address the following research questions:

Question 1 (Q1).   Is it possible to replicate the results of Hassenzahl, Platz, et al. (2000) with a real product from a domain that matches the experiences and concerns of the participants and provides a longer interaction time to build perceptions and judgment? A replication of key findings under different conditions would lend support to the reliability and validity of the research model.

The comparison of three different screen types—namely a standard CRT, a liquid crystal display (LCD), and a so-called virtual screen (VS; i.e., an image projected onto the desk by a ceiling-mounted projector)—provided the opportunity to study different but existing products that serve the identical purpose of displaying an image. An evaluation of real computer displays has more relevance than the evaluation of artificial software. Moreover, this type of evaluation involves different tasks and an extended interaction time relative to Hassenzahl, Platz, et al.

(2000). For these reasons, this study should broaden the experiential basis for the perception of EQ and HQ and the judgment of APPEAL.

Question 2 (Q2).   Do EQ, HQ, and APPEAL scales actually measure the intended constructs? Evidence of this would contribute to the validity of the scales and the research model.

To address Q2, an independent assessment of constructs similar to EQ, HQ, or APPEAL is necessary for comparison. In a usability study, Hassenzahl (2000) found that individuals who spent a higher proportion of task-completion time with usability problem-handling expended more *mental effort,* defined as the amount of energy an individual has to activate to meet the perceived task demands (Arnold, 1999). Individuals who experience many or hard-to-overcome usability problems or both (indicated by a higher rate of expended mental effort) should perceive the system as considerably less ergonomic and should rate the system as having less EQ. Furthermore, a higher expenditure of mental effort goes together with a reduced liking of a given system (Arnold, 1999). Given that liking and appeal are similar concepts, individuals with a higher expenditure of mental effort should rate the system to be less appealing (APPEAL). The perceptions of HQ should be independent from the expenditure of mental effort due to HQ's non-task-related character.

Question 3 (Q3).   Is it possible to provide an a priori prediction of the relative amount of user-perceived EQ and HQ per product from a given set of products? Moreover, can APPEAL be predicted on the basis of an averaging rule as assumed by Hassenzahl, Platz, et al. (2000)? A successful prediction of the screen types' EQ, HQ, and APPEAL would support the construct validity of the research model.

An informal inspection of the different screen types led to the following predictions. From an ergonomic perspective, the VS had some apparent drawbacks compared to the other two screen types. First, the image on the desk was relatively large, covering a significant amount of desk space. Users could perceive this as impractical for daily use. Second, the image was projected flatly onto the desk, making the required viewing angle awkward. In a normal sitting posture, users would experience image distortions. Most likely, they would compensate for this with an inconvenient, forward-bending seated posture. Based on these two drawbacks, it is likely that users would perceive the VS as having less EQ than the other two screens. Even though the CRT and LCD had different underlying technologies and screen sizes (19 in. [48.26 cm] and 15 in. [38.1 cm], respectively), the perceptible differences among them were minimal with regard to ergonomic variables such as reading speed, screen clarity, brightness, and contrast.

From a hedonic perspective, users should perceive the CRT as less hedonic than the other two screen types. Almost every computer user will have worked with and be familiar with a CRT. This is probably not true for the flat LCD and is surely not true for the VS. The CRT also lacks the potential gain in status derived from having a "smart" LCD or a "cool" VS sitting on one's desk. If this is true, users should perceive the CRT as having less HQ than the other two screens. If the APPEAL of the product is a consequence of perceptions of EQ and HQ, the LCD should have a higher APPEAL than the other two screens.

2. METHOD

2.1. Participants

Fifteen individuals (6 women and 9 men) participated in the study. Most participants were Siemens employees from Corporate Technology, Munich, Germany. All participants worked on visual display terminals regularly as a part of their job (e.g., secretaries, multimedia or Web designers, students). The sample's mean age was 35.4 years (ranging from 22 to 56). Computer expertise (measured with a five-item questionnaire) varied from *moderate* (10 participants) to *high* (5 participants).

2.2. Screen Types

The three tested screen types were (a) a standard screen with a 19-in. CRT, (Siemens MCM 1902); (b) a 15-in. LCD, (Siemens MCF 3811 TA); and (c) a 24-in. VS, that is, a projection of the screen from the ceiling flatly onto the user's desk (VS, Sharp Vision LC).

2.3. Measures

A semantic differential (see Hassenzahl, Platz, et al., 2000) was used to measure perceived EQ, perceived HQ, and APPEAL of the screen type. It consists of 23 seven-point scale items with bipolar verbal anchors (see Table 1). All verbal anchors were originally in German.

The Subjective Mental Effort Questionnaire (SMEQ; Arnold, 1999; German translation by Eilers, Nachreiner, & Hänecke, 1986; Zijlstra, 1993; Zijlstra & van Doorn, 1985) was applied to measure the expended mental effort. The SMEQ is a unidimensional rating scale ranging from 0 to 220. Different verbal anchors such as *hardly effortful* or *very effortful* facilitate the rating process.

2.4. Procedure

The study took place in the usability laboratory of Siemens Corporate Technology. Each participant came separately into the laboratory. After an introduction and instructions by the experimenter, each participant sat at a desk adapted from a typical office workplace. Each participant worked through three different tasks (a mahjongg game, an Internet search task, and a text editing task) with each screen type (CRT, LCD, and VS). The tasks covered many different aspects of working with a

Table 1: Bipolar Verbal Scale Anchors

Scale Item	Anchors	
EQ 1	Comprehensible	Incomprehensible
EQ 2	Supporting	Obstructing
EQ 3	Simple	Complex
EQ 4	Predictable	Unpredictable
EQ 5	Clear	Confusing
EQ 6	Trustworthy	Shady
EQ 7	Controllable	Uncontrollable
EQ 8	Familiar	Strange
HQ 1	Interesting	Boring
HQ 2	Costly	Cheap
HQ 3	Exciting	Dull
HQ 4	Exclusive	Standard
HQ 5	Impressive	Nondescript
HQ 6	Original	Ordinary
HQ 7	Innovative	Conservative
APPEAL 1	Pleasant	Unpleasant
APPEAL 2	Good	Bad
APPEAL 3	Aesthetic	Unaesthetic
APPEAL 4	Inviting	Rejecting
APPEAL 5	Attractive	Unattractive
APPEAL 6	Sympathetic	Unsympathetic
APPEAL 7	Motivating	Discouraging
APPEAL 8	Desirable	Undesirable

Note. Verbal anchors of the differential are translated from German. EQ = ergonomic quality; HQ = hedonic quality; APPEAL = appealingness.

computer. Participants used the screen types in random order. After finishing the three tasks with one screen type, each participant completed the SMEQ and the semantic differential, then repeated the procedure with the remaining screen types. Questionnaires concerning computer expertise and general demographics and a short interview completed the session. A whole session took from 2 to 3 hr; the interaction time with a single screen type ranged from 30 to 45 min.

3. RESULTS

3.1. Replication of Hassenzahl, Platz, et al. (2000; Q1)

Scale Validity: Principal Components Analysis

A slope analysis (Coovert & McNelis, 1988) of the eigenvalues from a PCA of the EQ and HQ items indicated a two-component solution. The varimax-rotated solution had a reasonably clear structure with HQ items loading on the first component and EQ items loading on the second component (Table 2, OVERALL). Together, the two components accounted for approximately 59% of the total variance. A Kaiser–Meyer–Olkin (Kaiser & Rice, 1974) measure of sample adequacy was .67, exceeding the required minimum of .5.

PCA requires a sample of independent measurements, but the present sample violates this requirement. Each participant evaluated each of the three screens and thus contributed three dependent measurements. This could reduce error variance, making the correlations in the matrix submitted to the PCA appear stronger than they actually were, which in turn may lead to an overestimation of explained variance or an unjustifiably clear component structure.

To check for this possible bias, I computed separate PCAs for each screen type. Because the model under test has two components (EQ and HQ), I restricted the solutions to an extraction of two components. Table 2 (CRT, LCD, VS) shows the resulting varimax-rotated solutions. The variance explained ranged from 50% to 60% of the total variance. Even in the worst case (LCD = 50.5%), the percentage of explained variance of the separate solution did not substantially differ from the OVERALL solution (OVERALL = 59.3%). Furthermore, each separate PCA more or less replicated the structure found in the OVERALL analysis, with HQ items consistently loading on one component and EQ items on the other. The only two obvious exceptions were the strong loading (.779) of the EQ item "supporting–obstructing" on the HQ component in the CRT analysis and the medium loading (.556) of the HQ item "innovative–conservative" on the EQ component of the VS analysis.

A PCA of the APPEAL items revealed only one major component, which accounted for about 68% of the variance. A Kaiser–Meyer–Olkin measure of sample adequacy was .83. Separate PCAs for each screen type all revealed one major component that accounted for 78% (CRT), 69% (LCD), and 54% (VS) of the variance, respectively.

Internal Consistency and Characteristics of the Single Scales

Table 3 summarizes the internal consistency (Cronbach's α) and general characteristics of each scale. Internal consistency was satisfactory. These results justify calculating EQ, HQ, and APPEAL values for each participant by averaging the single scale items.

Predicting the Participant's Judgment of Appealingness

Both EQ and HQ should affect the participant's judgment of APPEAL. To check this assumption, I conducted a regression analysis of EQ and HQ on APPEAL over all screen types (Table 4, OVERALL) and for each screen type separately (Table 4, CRT, LCD, VS).

Over all screen types, EQ and HQ effectively predicted APPEAL. Consistent with the assumptions, EQ and HQ contributed almost equally to APPEAL (see standardized regression coefficients, β). However, for the single screen types the re-

Table 2: Factorial Validity of Ergonomic Quality (EQ) and Hedonic Quality (HQ) Over All Screen Types and for Each Screen Type Separately

Principal Components With Varimax Rotation

Scale Item	OVERALL		CRT		LCD		VS	
	HQ	EQ	HQ	EQ	HQ	EQ	HQ	EQ
EQ								
Comprehensible–incomprehensible		.675		.667		.715		.534
Supporting–obstructing		.514	.779	.480	-.654			
Simple–complex		.759		.648		.754		.857
Predictable–unpredictable	-.483	.504				.716		.786
Clear–confusing		.740		.707		.717		.712
Trustworthy–shady		.802		.787		.798		.753
Controllable–uncontrollable		.887		.827		.892		.914
Familiar–strange	-.418	.636		.589	-.684			.701
HQ								
Interesting–boring	.857		.807			-.409	.652	
Costly–cheap	.519		.408		.609		.782	
Exciting–dull	.842		.728		.757		.847	
Exclusive–standard	.855		.826				.729	
Impressive–nondescript	.612		.594			-.659	.645	
Original–ordinary	.882		.889		.859		.755	
Innovative–conservative	.877		.884			-.511		.556
Eigenvalue	4.89	4.01	5.23	3.77	3.01	4.57	3.52	4.76
Variance explained (%)	32.57	26.76	34.88	25.15	20.04	30.45	23.48	31.71

Note. CRT = cathode-ray tube; LCD = liquid crystal display; VS = virtual screen. OVERALL *N* = 45 (15 participants × 3 prototypes). CRT, LCD, and VS *N* = 15; component loadings < .40 are omitted.

Table 3: Internal Consistency and Scale Characteristics

Scale	Cronbach's α	M	SD	Minimum	Maximum
EQ	.83	1.37	0.96	−1.75	2.88
HQ	.90	0.60	1.32	−2.43	2.86
APPEAL	.93	1.09	1.32	−2.63	3.00
SMEQ	X[a]	69.20	51.80	0	185

Note. $N = 45$ (15 participants × 3 screen types). EQ = ergonomic quality; HQ = hedonic quality; APPEAL = judgement of appeal; SMEQ = Subjective Mental Effort Questionnaire (Arnold, 1999).

[a]The SMEQ is a single item measurement tool; therefore, no Cronbach's α can be computed.

Table 4: Regression Analysis of EQ and HQ on APPEAL Over All Screen Types and for Each Screen Type Separately

Criterion	Adjusted R^2	Predictors	β	b	SE b	95% CI for b	
APPEAL (OVERALL)	.62**	EQ***	.62	.85	.13	.59	1.11
		HQ***	.61	.61	.09	.42	.80
APPEAL (CRT)	.70**	EQ**	.53	1.01	.29	.38	1.65
		HQ**	.55	.74	.20	.29	1.18
APPEAL (LCD)	.33*	EQ*	.71	.91	.31	.23	1.59
		HQ[a]	.47	.59	.30	−.08	1.25
APPEAL (VS)	.61**	EQ*	.40	.41	.17	.03	.79
		HQ**	.63	.79	.22	.32	1.26

Note. OVERALL $N = 45$ (15 participants × 3 screen types). EQ = ergonomic quality; HQ = hedonic quality; APPEAL = judgement of appealingness; CI = confidence interval; CRT = cathode-ray tube; LCD = liquid crystal display; VS = virtual screen. CRT, LCD, and VS $N = 15$.

[a]$p < .10$.

*$p < .05$. **$p < .01$. ***$p < .001$, two-tailed.

sults differed. Unlike the other screen types, LCD showed a low (but still significant) multiple correlation (adjusted R^2). For the CRT, EQ and HQ equally contributed to APPEAL, whereas for the LCD APPEAL was based more on EQ. For the VS, HQ had a more pronounced role. Although, the inspection of regression coefficients (*b*) implied differences in the contribution of either EQ or HQ to APPEAL, depending on the screen type, the 95% confidence intervals for the regression coefficient strongly overlapped. Thus, the data do not allow for a definite rejection of the equal contribution hypothesis.

3.2. Scale Validity: Correlation of SMEQ With EQ, HQ, and APPEAL (Q2)

It is reasonable to hypothesize that a high SMEQ value indicates usability problems experienced while performing the tasks. Participants should attribute a certain proportion of these problems to the EQ of the tested screen type, resulting in a perception of lower EQ. A highly significant negative correlation of SMEQ with EQ ($r =$ $-.61, p < .01$, two-tailed; $N = 45$) supported this assumption. Furthermore, experiencing usability problems (manifest in high SMEQ values) should lead to reduced product APPEAL, an assumption supported by a highly significant negative correlation of SMEQ and APPEAL ($r = -.58, p < .01$, two-tailed; $N = 45$). Finally, if HQ is a perception that is distinct and independent from perceptions of EQ, HQ should not correlate with experienced usability problems (indicated by high SMEQ values). In other words, experiencing usability problems should not affect a product's perceived HQ. Consistent with this assumption, the correlation between SMEQ and HQ was almost zero ($r = .01$, ns, two-tailed; $N = 45$).

3.3. A Priori Prediction of EQ, HQ, and APPEAL (Q3)

Figure 2 shows the mean scale values of EQ, HQ, and APPEAL for each screen type (CRT, LCD, VS).

I performed three separate repeated measurements analyses of variance (ANOVAs) with screen type (CRT, LCD, VS) as a within-subjects factor and each scale as a dependent variable (EQ, HQ, APPEAL). For each scale there was a significant main effect of screen type. Detailed analyses (by planned comparisons—repeated method) showed that participants perceived VS as less ergonomic (EQ) than LCD, with no differences between CRT and LCD. They perceived CRT as less hedonic (HQ) than LCD, with no differences between LCD and VS. Furthermore, they rated LCD as more appealing (APPEAL) than CRT and VS (see Table 5).

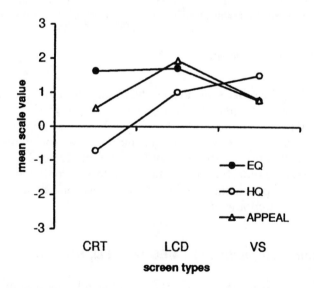

FIGURE 2   Mean scale values (EQ, HQ, APPEAL) for each screen type (CRT, LCD, VS).

**Table 5: Results of Repeated Measurements Analyses of Variance
for Each Scale (EQ, HQ, APPEAL) and Planned Comparisons**

Scale	Effect	F	df	Significance
EQ	Main effect	6.70	1.425^a	.011
	CRT vs. LCD	.17	1	ns
	LCD vs. VS	11.87	1	.004
HQ	Main effect	22.20	1.568^a	.000
	CRT vs. LCD	34.75	1	.000
	LCD vs. VS	2.56	1	ns
APPEAL	Main effect	5.70	1.452^a	.017
	CRT vs. LCD	7.75	1	.015
	LCD vs. VS	17.39	1	.001

Note. EQ = ergonomic quality; HQ = hedonic quality; APPEAL = judgement of appealingness; CRT = cathode-ray tube; LCD = liquid crystal display; VS = virtual screen.
[a]Greenhouse-Geisser corrected.

4. DISCUSSION

4.1. Replication of Hassenzahl, Platz, et al. (2000; Q1)

Q1 addressed the replication of the results of a previous study (Hassenzahl, Platz, et al., 2000) under very different conditions. In contrast to Hassenzahl, Platz, et al., this study included real products (computer screens) relevant to the participants, increasing the ecological validity of the results. Moreover, the three different tasks and the increased amount of interaction time gave participants the opportunity to build up a reasonable experiential basis for their perceptions and judgments.

PCAs

The PCAs indicated satisfactory component validity and internal consistency of EQ, HQ, and APPEAL. EQ and HQ are two distinctly perceived groups of quality dimensions respectively concerned with task-oriented or non-task-oriented quality aspects of the product. The results of the PCA of APPEAL items supported the notion of APPEAL being a unidimensional construct.

The OVERALL PCA of EQ and HQ items suffered from lack of independence of measurement. In an attempt to correct for this, I conducted separate analyses for each screen. However, with a sample size of 15 participants and 15 variables, the sample size to variables ratio of these analyses was only 1:1, which is lower than even the least conservative recommendations (e.g., Gorsuch, 1997, recommends a sample size to variables ratio of 3:1). PCAs with a low sample size to variables ratio tend to produce unstable component structures. For this reason, it is encouraging that each separate PCA resulted in components that explained a substantial portion of the total variance and, apart from minor exceptions, showed a consistent component structure.

The EQ items "predictable–unpredictable" and "familiar–strange" showed a moderate negative loading on the HQ component (–.483 and –.418, respectively). This indicates that a perception of a predictable, familiar product—two positive attributes from the usability perspective—goes together with a reduced perception of HQ. Carroll and Thomas (1988) pointed out that "ease of use" (i.e., EQ aspects) and "fun of use" (i.e., HQ aspects) may not necessarily complement each other. From a design perspective they may even be antagonistic, exclusive design goals. To make a product hedonic may only be possible by sacrificing part of its EQ and vice versa. Product designers should realize that HQ and EQ may be mutually exclusive concepts that they must bring into balance.

To conclude, the PCA presented herein is generally consistent with the preceding analyses. Nevertheless, it is important to replicate this analysis with a larger number of independent samples. A larger sample size would permit the use of confirmatory factor analysis or even structural equation methods to explicitly check the assumed relations among latent factors in the model.

Regression Analyses

The successful regression analysis of EQ and HQ on APPEAL over all screen types lent support to the notion that EQ and HQ perceptions equally contribute to the judgment of a product's APPEAL and tentatively supported the model of cognitive appraisal with its distinction between perception and evaluation. Subjective APPEAL appears to be a judgment that integrates different perceptions into a global construct. As proposed by information integration theory (Anderson, 1981; see also Dougherty & Shanteau, 1999), this might be the result of a cognitive averaging process. Such an averaging process implies, for example, that increased EQ can compensate for a lack of HQ with regard to the subjective evaluation of a product's APPEAL.

However, the regression analyses done separately for each screen type render a picture more complicated than a simple cognitive integration process. For the CRT, APPEAL was almost the numerical average of EQ and HQ. For the LCD, EQ affected the judgment of APPEAL more than HQ, and vice versa for the VS. This effect may be due to the participants' familiarity with the screen types. Participants had more experience with CRTs than with LCDs or VSs. Contrary to familiar objects, individuals may judge unfamiliar objects on the basis of particularly salient object features. Thus, the high-image quality of the LCD and the surprising novelty of the VS may lead to a higher weight of EQ and HQ, respectively.

4.2. Scale Validity: Correlation of SMEQ With EQ, HQ, and APPEAL (Q2)

Q2 addressed the topic of scale validity. I assessed construct validity of EQ, HQ, and APPEAL by correlating each scale with a measure of SMEQ. As expected, EQ and

APPEAL values were greater when participants experienced less severe usability problems (as indicated by a low SMEQ value). This lends support to the view that the EQ scale measures perceived (experienced) EQ and that the occurrence of usability problems during task execution lowers APPEAL. The lack of correlation of HQ with SMEQ points to the ability of the two scales (EQ and HQ) to discriminate between the distinct constructs.

However, validation of scales and their underlying constructs is far from being complete. The correspondence of EQ to other, more comprehensive usability evaluation questionnaires, such as the ISONORM 9241/10 (e.g., Prümper, 1999) or the IsoMetrics (Gediga, Hamborg, & Düntsch, 1999) deserves study. Second, experiments including more satisfaction and liking-oriented questionnaires such as the SUMI (Kirakowski & Corbett, 1993) and QUIS (e.g., Shneiderman, 1998) could elucidate the APPEAL construct. Regarding APPEAL, this construct integrates different product aspects (e.g., EQ, HQ), which is typical for judgments or evaluations. Also, APPEAL items are semantically evaluative (in contrast to the EQ and HQ items), but it is important to find a way to determine if EQ and HQ are truly more perceptual and APPEAL more evaluational in nature. Finally, it is important to develop methods for the explicit confirmation of the validity of the HQ construct. So far, I have only shown that HQ is different from EQ.

4.3. Prediction of EQ and HQ (Q3)

Q3 explored whether it is possible for an expert to use the research model to predict the relative EQ, HQ, and judgment of APPEAL of a given set of products. Detailed ANOVAs confirmed the a priori stated predictions.

However, there is one point to be discussed in more detail. In this study, the prediction of the relative amount of EQ and HQ per screen type was based on careful reasoning. From an interaction perspective, computer screens are fairly simple products that presumably serve only one practical purpose: to display information. Thus, display quality (readability) is one crucial criterion for predicting EQ. Moreover, the screen's intended context of use, on top of an office desk, is a well-known and understood context. Thus, the second criterion for EQ is the way the screen integrates into the workplace. Regarding HQ, it was reasonable to identify the novelty of the technology as the main source of HQ (see also Hassenzahl, Burmester, & Sandweg, 2000), for example, by the status gained from possessing a new technology in the social context of an organization.

The prediction of EQ and HQ heavily depends on a thorough analysis of the product and its context of use. A rational analysis might be adequate, given a simple technology in a well-understood context of use. For more complex products, used by experts in a special context of use, prediction of EQ is more or less impossible without reference to the wide array of analysis techniques used in the field of human–computer interaction (e.g., usability testing, expert evaluation methods, task and context analysis). The same is true for HQ, but here there are no specialized analysis techniques used in the industry. Therefore, it is important to develop appropriate analytical techniques to help product designers or usability engineers to gather *hedonic requirements* (i.e., user requirements addressing hedonic needs) that are appropriate in a certain context of use. The information gathered must support product designers or usability engineers in finding ways to fulfill hedonic requirements. Because EQ and HQ appear to be at least partially incompatible, finding trade-offs is also likely to be an important part of an appropriate analytical technique.

A first attempt to devise such a technique is the Structured Hierarchical Interviewing for Requirement Analysis (SHIRA; Hassenzahl, Wessler, & Hamborg, 2001; see also Hassenzahl, Beu, & Burmester, 2001). This interviewing technique seeks to contextualize abstract attributes such as original, innovative, and so forth. First, the interviewee explains what to his or her mind, for example, the term *originality* connotes for a specific product. Hereby, the transition from an abstract design goal (e.g., "to be original") to a number of concrete design goals (e.g., "to look different") is made. Second, the interviewee indicates ways to meet the concrete design goals (e.g., "do not use the typical Windows gray as the dominant color"). These solution proposals further typify the ideographic meaning of a concrete design goal. By aggregating the participants' ideographic views, a product's design space can be explored. If the attributes serving as the starting point of the interview are hedonic in nature, SHIRA has the potential to support the exploration of these and, thus, might help in gathering hedonic requirements.

To conclude, the successful prediction of EQ and HQ by referring to the research model clearly contributes to the validity of the model. However, it does not imply that a prediction of EQ and HQ based solely on the intuitions of product designers or usability engineers or both is likely to be successful without the use of appropriate analytical techniques.

5. HEDONIC QUALITY IN PRACTICE

To simultaneously consider EQ and HQ alters the focus of traditional usability engineering. In this section, I provide examples of how HQ affected some of the projects in which I was involved as a usability engineer.

In a project to design a telephone-based interface for a home automation system (Sandweg, Hassenzahl, & Kuhn, 2000), an evaluation of a first design revealed minor usability problems, an acceptable perceived EQ but a lowered HQ and APPEAL. We concluded that this lack of HQ (and APPEAL) was because of the design decision to use only spoken menus. A combination of spoken menus and nonspeech sounds was found to have an enriching effect on user–system interaction (Stevens, Brewster, Wright, & Edwards, 1994), which can be compared to the enriching effect of graphics on script (Brewster, 1998). Without the simultaneous consideration of perceived EQ and HQ, we might have missed the potential consequences of using speech sounds only on the APPEAL of the auditory interface.

HQ played an even more pronounced role in the redesign of a Web site we performed. A small-scale questionnaire study showed that the Web site's design lacked HQ, whereas EQ was perceived and experienced as very good. Participants described the design as visually dull and boring. This result provided an interesting and more or less new opportunity. Instead of ensuring a minimum of usability problems (as usability engineers usually do), we had to make the design more interesting, without compromising the good usability. We decided to perform a careful visual redesign, using mouse-over effects to provide a stronger impression of interactivity and some small animations to make transitions between pages more interesting. Unfortunately, no subsequent evaluation was performed, and thus, the question of whether HQ was increased by the redesign remains unanswered.

In a current project, we evaluated a redesigned software application for programming and configuring industrial automation solutions. The goals of the redesign were to increase usability—especially learnability—and to be recognized as innovative and original. Having both goals in mind, our client asked us to explicitly assess the EQ and HQ of the software to find appropriate trade-offs between usability and innovation. This involves, for example, the identification of "good" innovations, which enrich user experience without impairing usability.

These examples show the importance of taking HQ aspects into account. Had we not done so, we would have failed to recognize the potential improvement of the APPEAL of the home automation interface. The Web site's unappealing design would have gone unnoticed and a pure usability evaluation of the industrial automation software would have neglected important, market-relevant design goals.

Moreover, we have experienced an additional positive side effect since the introduction of HQ in our usability engineering practice: a better relationship with our clients' marketing departments. The relation between human factors and marketing is often described as chilly or distant (Nardi, 1995; but see Atyeo, Sidhu, Coyle, & Robinson, 1996, for a more positive perspective). This might be explained by the typically narrow perspective of usability engineering or human factors on product design. By solely focusing on usability and utility, some design goals (e.g., innovation) important to marketing (and to the user) are neglected.

6. CONCLUSION

The work discussed in this article provides a model for stimulating and guiding further research on product APPEAL. Many questions remain unanswered. For example, the relation of the APPEAL of a product to the consequences of this judgment is unknown. It would be interesting to study whether different distributions of EQ and HQ in comparable products (i.e., ergonomic-laden vs. hedonic-laden product types) may lead to different emotional reactions (e.g., satisfaction vs. joy). To give an example, although APPEAL of the CRT and the VS in this study is comparable, it stems from different sources: Participants perceived the CRT as mainly ergonomic (i.e., ergonomic laden), but they perceived the VS as mainly hedonic (i.e., hedonic laden). Although judged to be equally appealing, emotional or behavioral consequences might be quite different.

A second, most important topic of further research should be the analysis and transformation of hedonic requirements into actual product design. This means determining which features of a product intended for a certain context of use will induce the perception of HQ aspects. In other words, what makes a certain product appear to be interesting, innovative, or original? To answer this question, it will be necessary to complement the approach here with research that is more qualitative in nature.

With this article, I have attempted to put HQ—a quality aspect mostly irrelevant to the efficiency and effectiveness of a product—into perspective. Taking it into account when designing (or evaluating) a product allows software designers to go a step further, from simply making a product a useful tool to designing rich user experiences (Laurel, 1993). Whether a product's intended use is at work or at home, software designers should consider the gathering and analysis of hedonic requirements in addition to usability and functional requirements.

REFERENCES

Anderson, N. H. (1981). *Foundations of information integration theory.* New York: Academic Press.

Arnold, A. G. (1999). Mental effort and evaluation of user interfaces: A questionnaire approach. In *Proceedings of the HCI International 1999 Conference on Human Computer Interaction* (pp. 1003–1007). Mahwah, NJ: Lawrence Erlbaum Associates, Inc.

Atyeo, M., Sidhu, C., Coyle, G., & Robinson, S. (1996). Working with marketing. In *Proceedings of the CHI 1996 Conference Companion on Human Factors in Computing Systems: Common ground* (pp. 313–314). New York: ACM, Addison-Wesley.

Bevan, N. (1995). Usability is quality of use. In *Proceedings of the HCI International 1995 Conference on Human Computer Interaction* (pp. 349–354). Mahwah, NJ: Lawrence Erlbaum Associates, Inc.

Bevan, N., & Macleod, M. (1994). Usability measurement in context. *Behaviour & Information Technology, 13,* 132–145.

Brewster, S. A. (1998). Using nonspeech sounds to provide navigation cues. *Transactions on Computer-Human Interaction, 5,* 224–259.

Carroll, J. M., & Thomas, J. C. (1988). Fun. *SIGCHI Bulletin, 19*(3), 21–24.

Coovert, M. D., & McNelis, K. (1988). Determining the number of common factors in factor analysis: A review and program. *Educational and Psychological Measurement, 48,* 687–693.

Doll, W. J., & Torkzadeh, G. (1988). The measurement of end user computing satisfaction. *MIS Quarterly, 12,* 259–274.

Dougherty, M. R. P., & Shanteau, J. (1999). Averaging expectancies and perceptual experiences in the assessment of quality. *Acta Psychologica, 101,* 49–67.

Eilers, K., Nachreiner, F., & Hänecke, K. (1986). Entwicklung und Überprüfung einer Skala zur Erfassung subjektiv erlebter Antrengung [Development and evaluation of a questionnaire to assess subjectively experienced effort]. *Zeitschrift für Arbeitswissenschaft, 40,* 215–224.

Ekman, P., & Davidson, R. J. (1994). *The nature of emotion.* New York: Oxford University Press.

Gediga, G., Hamborg, K.-C., & Düntsch, I. (1999). The IsoMetrics usability inventory: An operationalization of ISO 9241–10 supporting summative and formative evaluation of software systems. *Behaviour & Information Technology, 18,* 151–164.

Gorsuch, R. L. (1997). Exploratory factor analysis: Its role in item analysis. *Journal of Personality Assessment, 68,* 532–560.

Harrison, A. W., & Rainer, R. K. (1996). A general measure of user computing satisfaction. *Computers in Human Behavior, 12*(1), 79–92.

Hassenzahl, M. (2000). Prioritising usability problems: Data-driven and judgement-driven severity estimates. *Behaviour & Information Technology, 19,* 29–42.

Hassenzahl, M., Beu, A., & Burmester, M. (2001). Engineering joy. *IEEE Software, 1&2,* 70–76.

Hassenzahl, M., Burmester, M., & Sandweg, N. (2000). Perceived novelty of functions—A source of hedonic quality. *Interfaces, 42,* 11.

Hassenzahl, M., Platz, A., Burmester, M., & Lehner, K. (2000). Hedonic and ergonomic quality aspects determine a software's appeal. In *Proceedings of the CHI 2000 Conference on Human Factors in Computing Systems* (pp. 201–208). New York: ACM, Addison-Wesley.

Hassenzahl, M., Wessler, R., & Hamborg, K.-C. (2001). Exploring and understanding product qualities that users desire. In *Proceedings of the IHM/HCI Conference on human-computer interaction, Volume 2* (pp. 95–96). Toulaise, France: Cépadinès.

Igbaria, M., Schiffman, S. J., & Wieckowski, T. J. (1994). The respective roles of perceived usefulness and perceived fun in the acceptance of microcomputer technology. *Behaviour & Information Technology, 13,* 349–361.

ISO. (1998). *Ergonomic requirements for office work with visual display terminals (VDTs)—Part 11: Guidance on usability* (ISO No. 9241). Geneva, Switzerland: Author.

Kaiser, H. F., & Rice, J. (1974). Little Jiffy, Mark IV. *Educational and Psychological Measurement, 34,* 111–117.

Kirakowski, J., & Corbett, M. (1993). SUMI: The Software Measurement Inventory. *British Journal of Educational Technology, 24,* 210–212.

Kurosu, M., & Kashimura, K. (1995). Apparent usability vs. inherent usability. In *Proceedings of the CHI 1995 Conference Companion on Human Factors in Computing Systems* (pp. 292–293). New York: ACM, Addison-Wesley.

Laurel, B. (1993). *Computers as theatre.* Reading, MA: Addison-Wesley.

Leventhal, L., Teasley, B., Blumenthal, B., Instone, K., Stone, D., & Donskoy, M. V. (1996). Assessing user interfaces for diverse user groups: Evaluation strategies and defining characteristics. *Behaviour & Information Technology, 15,* 127–137.

Mundorf, N., Westin, S., & Dholakia, N. (1993). Effects of hedonic components and user's gender on the acceptance of screen-based information services. *Behaviour & Information Technology, 12,* 293–303.

Nardi, B. (1995). Some reflections on scenarios. In J. M. Carroll (Ed.), *Scenario-based design: Envisioning work and technology in system development* (pp. 397–399). New York: Wiley.

Ortony, A., Clore, G. L., & Collins, A. (1988). *The cognitive structure of emotions.* Cambridge, MA: Cambridge University Press.

Prümper, J. (1999). Test IT: ISONORM 9241/10. In *Proceedings of the HCI International 1999 Conference on Human Computer Interaction* (pp. 1028–1032). Mahwah, NJ: Lawrence Erlbaum Associates, Inc.

Sandweg, N., Hassenzahl, M., & Kuhn, K. (2000). Designing a telephone-based interface for a home automation system. *International Journal of Human–Computer Interaction, 12,* 401–414.

Shneiderman, B. (1998). *Designing the user interface: Strategies for effective human–computer interaction* (3rd ed.). Reading, MA: Addison-Wesley.

Stevens, R. D., Brewster, S. A., Wright, P. C., & Edwards, A. D. N. (1994). Providing an audio glance at algebra for blind readers. In G. Kramer & S. Smith (Eds.), *Proceedings of ICAD '94* (pp. 21–30). Santa Fe, NM: Santa Fe Institute, Addison-Wesley.

Zijlstra, R. (1993). *Efficiency in work behaviour: A design approach for modern tools.* Delft, The Netherlands: Delft University Press.

Zijlstra, R., & van Doorn, L. (1985). The construction of a scale to measure subjective effort. Delft, The Netherlands: Delft University of Technology, Department of Philosophy and Social Sciences.

INTERNATIONAL JOURNAL OF HUMAN–COMPUTER INTERACTION, *13*(4), 501
Copyright © 2001, Lawrence Erlbaum Associates, Inc.

Schedule of Events
2002

March 16 – 19   **Nice, France**
Human Factors & the Web: 1st International Conference
URL: http://www.unice.fr/ihfweb/intconf/

April 20 – 25   **Minneapolis, Minnesota, USA**
CHI 2002: Conference on Human Factors in Computing Systems
URL: http://www.acm.org/sigchi/chi2002/indexcopy.html

May 22 – 25   **Berchtesgaden, Germany**
WWDU 2002: World wide work with information and communications technology
URL: http://wwdu.org/2002

July 8-10   **Edinburgh, Scotland**
ASSETS 2002: The 5th International ACM SIGCAPH Conference on Assistive Technologies
URL: http://www.acm.org/sigcaph/assets02/

INTERNATIONAL JOURNAL OF HUMAN–COMPUTER INTERACTION, *13*(4), 503
Copyright © 2001, Lawrence Erlbaum Associates, Inc.

Author Index to Volume 13

INTERNATIONAL JOURNAL OF HUMAN–COMPUTER INTERACTION, *13*(4), 505–508
Copyright © 2001, Lawrence Erlbaum Associates, Inc.

Content Index to Volume 13

Number 1, 2001

Number 2, 2001

Contents:

Articles:

Number 3, 2001

Articles:

Number 4, 2001

Contents:

Articles: